CO-ATR-895

Hot Topics

Hot Topics
Everyday Environmental Concerns

S.A. Abbasi
P. Krishnakumari
and
F.I. Khan

OXFORD
UNIVERSITY PRESS

OXFORD
UNIVERSITY PRESS

YMCA Library Building, Jai Singh Road, New Delhi 110 001

Oxford University Press is a department of the University of Oxford. It furthers the
University's objective of excellence in research, scholarship, and education
by publishing worldwide in

Oxford   New York

Athens   Auckland   Bangkok   Bogota   Buenos Aires   Calcutta
Cape Town   Chennai   Dar es Salaam   Delhi   Florence   Hong Kong   Istanbul
Karachi   Kuala Lumpur   Madrid   Melbourne   Mexico City   Mumbai
Nairobi   Paris   Sao Paulo   Singapore   Taipei   Tokyo   Toronto   Warsaw
with associated companies in   Berlin   Ibadan

Oxford is a registered trade mark of Oxford University Press
in the UK and in certain other countries

Published in India
By Oxford University Press, New Delhi

ISBN   019   564 5340

Typeset in Palatino
by Eleven Arts, Keshav Puram, Delhi 110 035
Printed by Rashtriya Printers Pvt. Ltd., Delhi 110 032
Published by Manzar Khan, Oxford University Press
YMCA Library Building, Jai Singh Road, New Delhi 110 001

This book is dedicated to

Mani Shankar Aiyar and his secular initiative in dousing fires
of a very destructive kind

—Abbasi

My late aunt, Professor A. Lakshmikutty Pisharasiar

—Krishnakumari

My parents, Shri Irshad Ali Khan Kakarzai and
Smt Mahmooda Begam

—Faisal

Contents

Figures

Tables

Foreword

A l-Biruni, one of the most famous scientists of all time, had classified human diseases into two broad categories—the diseases of the rich and those of the poor. Environmentalists today can take a cue from Al-Biruni and put environmental 'diseases' also in two categories: the ones generated by the rich and the ones generated by the poor. Problems like global warming, ozone hole, and acid rain would fall in the former category and the ones such as sanitary pollution in the latter. But one essential difference between human diseases and environmental diseases is that the former do not necessarily cut across economic strata whereas the latter almost always do. In fact it is the poor who suffer a lot more from the environmental problems created by the rich than vice versa. It is also pertinent to mention that of the five major environmental problems discussed in this book, four are generated by the activities of the rich.

Irrespective of how they originate, the five environmental problems are unarguably the most important issues of global concern we confront today. This book, though written in India, has an appeal and a relevance which transcends geographical boundaries.

Whenever we talk of these 'hot' problems we tend to curse industrialization and the resultant developments in transportation, refrigeration, air-conditioning, and in the manufacture of all kinds of chemicals. There is always this talk of exposure to carcinogens and mutagens by xenobiotics; the talk of higher incidence of heart diseases, cancer, and other evils. As if we are suffering because of technological progress and would have been better off without it. What is almost always overlooked is the fact that life expectancy all over the world has gone up. More and more diseases which earlier used to be fatal are not so now. This is the positive side of technical progress.

What is also overlooked is that technological progress has brought into our lives so much more colour, energy, and vibrance that we would not like to pass a single day without these comforts. But there is a tendency to take all this for granted and focus only on the harmful fall-outs.

What is beyond argument is the fact that environmental pollution does harm the quality of life we lead. But for pollution, we would have been eating the cake of technological progress and having it too! This makes all initiatives for solving these problems timely and crucial. *Hot Topics* contributes a great deal towards such initiatives.

A major attribute of *Hot Topics* is that it has treated the five most daunting and pervasive environmental problems in a balanced manner. There is no premediated slant 'for' or 'against'; nor is there any attempt at exaggerating the magnitude of the problems.

Tariq M. Babtain
General Director
Saudi Environmental Works Ltd.
Al-Khobar

Preface

Ecology is one of the classical branches of the life sciences, nearly as old as taxonomy, physiology, and anatomy. Even the sub-branches of ecology—such as limnology, marine ecology, and forest ecology—are over 100 years old. Yet ecology remained a field pursued, discussed, and reported only by trained scientists till the late 1960s. Then came the club of Rome report closely followed by Dennis Meadow's book *Limits to Growth*. These two, especially Meadow's blockbuster, shook the world as few other scientific works have, ever since Einstien's theory of relativity was 'proved' in 1919. These writings predicted that the world will literally go up in smog by the mid-twenty-first century, making humans extinct, if environmental pollution and resource consumption were not curbed on a war-footing. All of a sudden, terms such as 'environment', 'pollution', 'ecosystem', 'holistic view', 'degradation' gained wide currency and came to be used by everyone in day-to-day conversation.

And ecology came riding on the crest of a giant wave, spreading to all the nooks and corners of the world!

If ecology is an old term which has recently become fashionable, environmental science *is* a relatively new term. In its literal meaning and implication, environmental science covers a larger domain than ecology—according to some much too large for comfort! Anything and everything under the sun (and beyond it) can be brought under the purview of environmental science, and attempts to specify the boundaries of this field of knowledge quite often end up covering the whole universe.

But whichever definition of environmental science one may choose to accept, there are a few topics of global concern which would invariably be encompassed by it. It is to these 'hot topics' that the present book has been devoted.

We have used the term 'hot topics' in the literal as well as metaphorical sense. If global warming is all about increasing temperatures, the ozone hole is about the passage of extra ultraviolet radiation through the thinning blanket of ozone to scorch our skin. We may have acid rain pouring down and causing skin-burns, or hazardous nuclear wastes generating 'heat' of a very special and frightening kind. Even if we may escape all these scourges, we may be caught napping when a refinery or a tanker goes aflame one fine morning as it did in Vishakhapattnam recently. And we may cop the 'heat' generated by typhoid, cholera, malaria and other such diseases if the water we drink is as questionably disinfected as most of our water supply is.

Each of these 'hot' topics is of vital importance all over the globe, and many of them are interrelated. For example, several gases which cause global warming also generate acid rain. Some of them worsen, and some help, the cause of protecting the ozone layer. Handling of hazardous substances is intimately linked with the risk of industrial accidents...and so on. In this book we have addressed these hot topics together to give an indication of their individual contributions to the environmental problems of our times as well as their collective impact.

S.A. Abbasi
P. Krishnakumari

Pondicherry   F.I. Khan

1

The Greenhouse Effect

I n the last two decades, the term 'environment' has gained enormous importance and almost everyone is aware of the term. Along with it, terms like 'ozone depletion', 'greenhouse effect' and 'acid rain' have also surfaced and are often the topic of discussion among the socially and educationally advanced. But what exactly do these terms mean? How real are the threats posed by these changes and processes? To begin with let us take a look at *the greenhouse effect*.

What is the greenhouse effect?

As the name suggests, the term had its origin from the practice in cold countries of encasing vegetation in glass chambers to protect them from frost. It was observed that there was a continued rise in temperature in such chambers even when the outside temperatures remained low. This enabled the warming up of vegetation inside the chamber, resulting in good plant growth.

Figure 1.1 illustrates what happens in a typical greenhouse. The transparent glass roof and walls of the greenhouse allow the sun's rays to pass through and strike the ground (surface of the chamber). The reflected radiation is of longer wavelength than the incident radiation. A significant portion of the former is absorbed by the glass. As long wavelength radiation (infra-red radiation) generate heat, this results in a rise in temperature inside the greenhouse.

Figure 1.2 shows that an effect similar to the greenhouse effect is responsible for keeping the earth's surface warmer than it

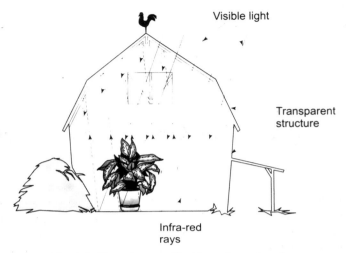

Visible light

Transparent
structure

Infra-red
rays

Fig. 1.1 Greenhouse effect in a glass chamber

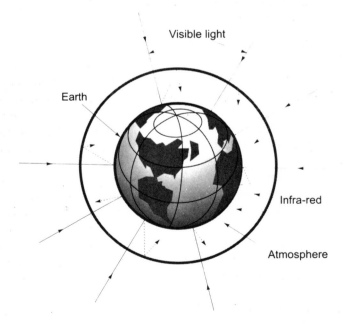

Visible light

Earth

Infra-red

Atmosphere

Fig. 1.2 Greenhouse-like effect on the earth

would otherwise be. The sun's rays strike the surface of the earth and the long wavelength radiation emitted by the earth are absorbed by atmospheric gases, thereby contributing to the rise in temperature.

The possibility that the absorption of long wavelength radiation by atmospheric gases would influence ground temperature, was recognized by Fourier as early as 1827. Fourier maintained that the atmosphere acts like the glass of a greenhouse by letting through the incident light rays of the sun and retaining the infra-red rays, which are reflected back to the ground. This greenhouse effect warms the lower atmosphere.

If the atmosphere was transparent to the outgoing long wavelength radiation emanating from the earth's surface, the mean equilibrium temperature of the earth's surface would be considerably lower and probably below the freezing point of water. According to one estimate, in the absence of natural concentrations of greenhouse gases, the average temperature of the earth's surface would be -19°C instead of the present value of 15°C and the earth would be a frozen lifeless planet. Working along these lines, the Swedish chemist Svante Arrhenius was able to show, in 1896, that it was the carbon dioxide (CO_2)present in the earth's atmosphere which helped the atmosphere retain the long wavelength radiation and thus warm up the earth.

The earth's radiation balance

Scientists, until 1896, had been unable to explain how the earth's atmosphere could maintain the planet's relatively warm temperature, when oxygen and nitrogen, which constitute 99 per cent of the atmosphere, do not absorb heat from the infra-red radiation emitted from the earth back into space. The emission spectra of the sun roughly resembles that of a blackbody, radiating at a temperature of 6000°K. In the visible portion of the spectrum (0.4 to 0.7 μm wavelength range), where the maximum influx of solar energy takes place, the radiation can penetrate, almost without loss, down to the earth's surface except where clouds are present. High in the atmosphere ordinary oxygen (O_2) and ozone (O_3) molecules absorb an estimated 1–3 per cent of the incoming radiation. The absorption occurs in the ultraviolet portion of the spectrum and effectively limits the penetrating radiation to wavelengths longer than 0.3 μm. Although this effect is relatively small, it is important because it is the main source of energy for circulation of gases in the atmosphere above 30 kilometres. Moreover the absorption at these levels shields the biosphere from the damaging effects of ultraviolet radiation. In spite of certain long term climatic changes, climatological records do not show an appreciable net heating of the earth and its atmosphere. Therefore, the earth must be radiating as much as the radiation absorbed. A characteristic shift to longer wavelengths does take place. However, since the earth radiates at an effective blackbody temperature of 255°K, a very low temperature compared to the sun's blackbody temperature of 6000°K, the earth's emission occurs over a broad range of wavelengths from 2 to 40 microns with a flat maximum at about 12 microns. In this range the atmosphere is no longer transparent. Arrhenius discovered that carbon dioxide, which makes up only a tiny fraction of the atmosphere, could trap enough of the escaping heat (wavelengths of 12–16.3 microns) to warm up the surface of the planet. Furthermore he realized that the burning of coal, oil and natural gases was raising the concentrations of carbon dioxide and he predicted that a doubling of the gas could warm the planet by more than 10°F, a prediction that is considered reasonable by modern-day scientists.

What are the greenhouse gases?

The greenhouse gases of greatest concern are carbon dioxide, water vapour, methane, chlorofluorocarbons, nitrogen oxides, and tropospheric ozone.

Carbon Dioxide: (CO_2) In 1958, Charles Keeling and Roger Revalle of the Scripps Institution of Oceanography at La Jolla, USA, studied the atmosphere's carbon dioxide concentrations. On the slopes of Mauna Loa in Hawaii, far from any sources of industrial pollution, Keeling was able to measure the subtle, seasonal ups and downs in the concentration of the CO_2 as the plants in the Northern Hemisphere varied their consumption and release of the gas. In addition to this annual cycle, he documented a steady rise in carbon dioxide due to human activities from 315 parts

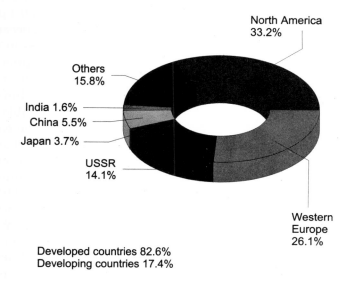

North America
33.2%

Others
15.8%

India 1.6%
China 5.5%
Japan 3.7%

USSR
14.1%

Western
Europe
26.1%

Developed countries 82.6%
Developing countries 17.4%

Fig. 1.3 Contributions to increases in atmospheric carbon dioxide concentration (1800–1988)

per million (ppm) thirty years ago to 350 ppm today. Using these data as well as evidence from tree rings and ice cores, climatologists estimate that the level of carbon dioxide before the dawn of the Industrial Revolution was about 280 ppm. Humans have already increased the levels of CO_2 by 25 per cent and are expected to double the CO_2 levels by the year 2075. Figure 1.3 depicts the contributions of different nations to the increase in global atmospheric carbon dioxide concentrations from 1800 to 1988.

The estimate of CO_2 emissions (global) from 1950 to 1988 are provided in table 1.1, and those for India in table 1.2. The estimated sources and sinks of CO_2 are given in table 1.3.

Table 1.1 CO_2 emission estimates[a] (global) (1950–88)

Year	Total	Fuels			Cement	Gas flaring	Per capita[b]
		Gaseous	Liquid	Solid			
1950	1638	97	423	1077	18	23	0.7
1951	1775	115	479	1137	20	24	0.7
1952	1803	124	504	1127	22	26	0.7
1953	1848	131	533	1132	24	27	0.7
1954	1871	138	557	1123	27	27	0.7
1955	2050	150	625	1215	30	31	0.7
1956	2185	161	679	1281	32	32	0.8
1957	2278	178	714	1317	34	35	0.8
1958	2338	192	732	1344	36	35	0.8
1959	2471	214	790	1390	40	36	0.8
1960	2586	235	850	1419	43	39	0.9
1961	2602	254	905	1356	45	42	0.9
1962	2708	277	981	1358	59	44	0.9
1963	2855	300	1053	1404	51	47	0.9
1964	3016	328	1138	1442	57	51	0.9
1965	3154	351	1221	1468	59	55	1.0
1966	3314	380	1325	1485	63	60	1.0
1967	3420	410	1424	1455	65	66	1.0
1968	3596	445	1552	1456	70	73	1.0
1969	3809	487	1674	1494	74	80	1.1
1970	4090	515	1838	1571	78	87	1.1
1971	4241	553	1946	1571	84	88	1.1
1972	4409	582	2056	1587	89	95	1.2
1973	4647	607	2240	1594	95	110	1.2
1974	4655	616	2214	1591	96	108	1.2
1975	4628	620	2131	1686	95	95	1.1
1976	4894	644	2313	1723	103	111	1.2
1977	5034	645	2390	1786	108	105	1.2
1978	5082	673	2383	1802	116	107	1.2
1979	5365	713	2535	1899	119	100	1.2
1980	5263	724	2409	1921	120	89	1.2
1981	5129	734	2272	1930	121	72	1.1
1982	5093	732	2178	1993	121	70	1.1

Contd.

Table 1.1 (Contd.)

Year	Total	Fuels			Cement	Gas flaring	Per capita[b]
		Gaseous	Liquid	Solid			
1983	5084	735	2163	1998	125	63	1.1
1984	5260	796	2191	2088	128	57	1.1
1985	5379	826	2172	2196	130	55	1.1
1986	5561	842	2277	2253	136	53	1.1
1987	5680	888	2290	2313	142	48	1.1
1988	5893	919	2392	2385	150	48	1.2

a   emission estimates rounded off and expressed in million tonnes of carbon.
b   per capita estimates rounded off and expressed in tonnes of carbon
Source: Oak Ridge National Laboratory, 1990, *Trends '90: A Compendium of Data on Global Change*, Oak Ridge, Tennessee, USA: The CO_2 Information Analysis Centre

Table 1.2 CO_2 emission estimates[a] (India) (1950–88)

Year	Total	Fuels			Cement	Gas flaring	Per capita[b]
		Gaseous	Liquid	Solid			
1950	18.4	16.0	2.0	0.0	0.4	0.0	0.1
1951	19.2	16.6	2.1	0.0	0.4	0.0	0.1
1952	20.3	17.7	2.1	0.0	0.5	0.0	0.1
1953	20.6	18.0	2.1	0.0	0.5	0.0	0.1
1954	21.7	18.5	2.6	0.0	0.6	0.0	0.1
1955	23.5	19.4	3.5	0.0	0.6	0.0	0.1
1956	24.4	19.9	3.8	0.0	0.7	0.0	0.1
1957	27.3	22.2	4.3	0.0	0.8	0.0	0.1
1958	28.6	23.1	4.6	0.0	0.8	0.0	0.1
1959	30.2	24.3	5.0	0.0	0.9	0.0	0.1
1960	33.2	26.8	5.3	0.0	1.1	0.0	0.1
1961	35.9	29.0	5.8	0.0	1.1	0.0	0.1
1962	39.5	31.5	6.8	0.0	1.2	0.0	0.1

Contd.

Table 1.2 (Contd.)

Year	Total	Fuels			Cement	Gas flaring	Per capita[b]
		Gaseous	Liquid	Solid			
1963	42.4	33.7	7.4	0.0	1.3	0.0	0.1
1964	41.4	32.3	7.7	0.0	1.3	0.1	0.1
1965	45.6	35.5	8.4	0.1	1.4	0.2	0.1
1966	47.2	35.7	9.7	0.1	1.5	0.2	0.1
1967	47.4	36.3	8.9	0.1	1.6	0.4	0.1
1968	51.5	37.7	11.5	0.2	1.6	0.4	0.1
1969	52.4	37.6	12.4	0.3	1.8	0.3	0.1
1970	53.3	37.7	13.1	0.3	1.8	0.4	0.1
1971	56.2	38.7	14.1	0.3	2.0	0.4	0.1
1972	59.4	41.0	15.6	0.3	2.1	0.3	0.1
1973	61.1	42.0	16.4	0.3	2.0	0.4	0.1
1974	63.6	44.0	16.4	0.4	1.9	0.5	0.1
1975	69.1	49.2	16.7	0.5	2.2	0.5	0.1
1976	72.2	51.5	17.1	0.6	2.5	0.5	0.1
1977	86.5	64.2	18.4	0.7	2.6	0.6	0.1
1978	87.0	62.9	20.2	0.7	2.7	0.5	0.1
1979	90.8	64.8	22.2	0.8	2.5	0.6	0.1
1980	95.5	69.7	22.4	0.7	2.4	0.3	0.1
1981	102.7	74.1	24.1	0.8	2.8	0.9	0.1
1982	109.2	78.4	25.5	1.2	3.1	1.0	0.2
1983	118.5	85.7	26.7	1.4	3.4	1.2	0.2
1984	122.5	86.8	28.6	1.7	3.9	1.4	0.2
1985	134.3	94.8	31.4	2.0	4.5	1.6	0.2
1986	143.6	101.5	32.9	2.8	5.0	1.4	0.2
1987	152.1	108.5	33.7	3.2	5.0	1.7	0.2
1988	163.8	117.4	35.3	3.5	5.5	2.0	0.2

a   emission estimates rounded off and expressed in million tonnes of carbon

b   per capita estimates rounded off and expressed in tonnes of carbon.

Source: Oak Ridge National Laboratory, 1990, *Trends '90: A Compendium of Data on Global Change*, Oak Ridge, Tennessee, USA: The CO_2 Information Analysis Centre

Table 1.3 Estimated sources and sinks of CO_2

Sources/Sinks	Range (TgC/year)
Sources	
Ocean	102700–106500
Land	8700–120000
Fossil fuel	4500–5500
Land use conversion	0–2600
Sinks	
Oceans	106000–108000
Land	100000–140000

Source: A Primer on Energy, Dept. of Energy, USA [Tg=Teragrams=10^{12}g]

The Carbon Cycle: The carbon cycle in the biosphere (figure 1.4) is fundamentally an overall global interference of living organisms and their physical and chemical environment. Carbon dioxide is the most abundant and the *single most* important greenhouse gas in the atmosphere. Its concentration has increased by about 25 per cent since the industrial revolution. Detailed measurements since 1958 show an increase from 315 to 350 parts per million by volume. Carbon dioxide concentrations are currently increasing at a rate of about 0.4 per cent per year, which is responsible for about half of the current increase in global warming caused by greenhouse gases. Deforestation and consumption of fossil fuels have both contributed to this rise. Current emissions are estimated at 5.5 billion tonnes of carbon from fossil fuel combustion and 0.4–2.6 billion tonnes of carbon from deforestation. This carbon dioxide remains in the atmosphere or is absorbed by the oceans. Even though only half of the current emissions remain in the atmosphere, available models of CO_2 uptake by the ocean suggest that substantially more than a 50 per cent cut in the emissions is required to stabilize concentrations at current levels. Figure 1.4 makes the carbon cycle look simple but in reality it is enormously complex. CO_2 levels vary in the atmosphere not only diurnally or seasonally but spatial variations—latitudinal and altitudinal— have also been observed. The *average* global CO_2 concentration is about 320 ppm by volume. Paleoclimatological studies show

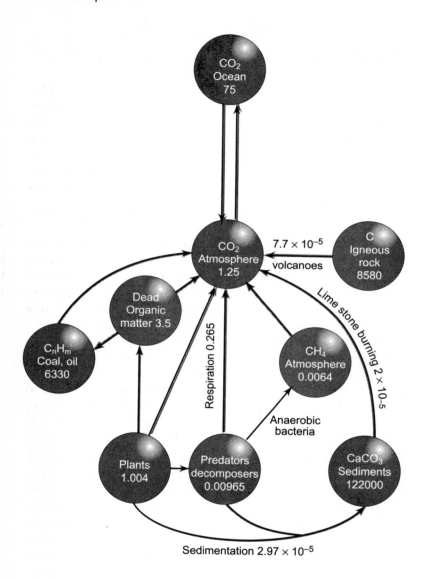

Note: The numbers indicate giga tonnes

Fig. 1.4 The Carbon Cycle (Jogensen 1983)

that during the last glacial maximum, the CO_2 concentration was only 180–220 ppm (Golitsyn 1993). But one of the most unexpected discoveries of recent years was that the CO_2 concentration could change by a factor of 1.5 during a time span of only a few hundred years. The fact that the changes are real is confirmed by the observation that they are accompanied by corresponding changes in the $^{18}O/^{16}O$ ratio oxygen istopes; (isotopes are different forms of an element that differ in relative atomic mass and in their nuclear but not chemical properties), which reflects the decrease in temperature in this period, evidently due to the decrease of atmospheric greenhouse effect. This fact shows that we are still far from understanding many important factors determining the carbon cycle in nature.

$\delta^{13}C$ (δ = delta) records preserved in the peat bogs at high altitudes of the Nilgiris are also good indicators of quaternary climatic conditions. $\delta^{13}C$ is the ratio of ^{13}C isotope to ^{12}C form, which varies in the different types of vegetation. The C_3 plants (wheat, rice, legumes, oilseeds and cotton) have $\delta^{13}C$ values in the range of 26–28 per cent, while C_4 plants (sugarcane, sorghum and maize) assimilate ^{13}C at different levels giving a range of 11–13 per cent. The $\delta^{13}C$ records give an indication of the past vegetation trends, whereby the low values signify a C_4 dominated community which in turn imply a lowered atmospheric temperature and low soil-moisture (Sukumar et al. 1993).

The forests contain about 400–500 billion tonnes of carbon or roughly two-third of the amount present in the atmosphere which is about 700 billion tonnes. The northern temperate forests also sequester a good amount of excess atmospheric CO_2 (Sedjo 1992). The role of biomass in sequestering atmospheric carbon dioxide has been detailed by Rossillo et al. (1992). The perfect balance of the carbon cycle and the role of ocean as a major shock absorber warrants detailed study.

Methane (CH_4): Fewer studies have been conducted on the impact of methane as a greenhouse gas compared to CO_2 due to the greater uncertainties about the sources and sinks of methane (Rotmans et al. 1992). Table 1.4 presents the estimated sources and sinks of methane. Apart from its role in global warming, methane can also affect the tropospheric ozone

Table 1.4 Global methane emission estimates

Sources/sinks	Best estimate (10^6 ta^{-1}) (10^6 ta^{-1})*	Range
Sources		
Natural wetlands	115	100–200
Rice paddies	110	25–170
Enteric fermentation (animals)	80	65–100
Gas drilling, venting, transmission	45	25–50
Biomass burning	40	20–80
Termites	40	10–100
Landfills	40	20–70
Coal mining	35	19–50
Oceans	10	5–20
Fresh waters	5	1–25
CH_4 hydrate destabilisation	5	0–100
Sinks		
Removal by soils	30	15–45
Reaction with OH	500	400–600

Source: Houghton et al. 1990
*ta^{-1}: tonnes per annum

distribution. A model named IMAGE (Integrated Model to Assess Greenhouse Effects) predicts that a 10 per cent cut in CH_4 emissions, besides stabilizing carbon dioxide emissions by the year 2025, can lead to stable CH_4 levels in the atmosphere. CH_4 is currently increasing at a rate of 1 per cent per year and is responsible for about 20 per cent of current increases in the commitment to global warming. Scientists from developed countries have blamed paddy fields, existing predominantly in tropical developing countries yielding about 90 per cent of global rice production, for contributing vast quantities of methane to the global methane flux. These allegations have been hotly contested by the developing countries and this has persistently been a bone of contention in the North–South dialogue on global warming. The extent of divergence of views on this subject can be guaged from the widely differing estimates of methane contribution from rice paddies. For example an Indian scientist Sinha has claimed that the estimates of IPCC

(Inter Governmental Panel on Climate Change) on methane production in developing countries are as much as 15 times higher than actual (*Down to Earth*, 1995).

The emissions from landfills and coal seams also contribute to the methane levels all over the world. CO_2 and CH_4 are known to be produced during bacterial decomposition of flooded peat and forest biomass. Hydroelectric reservoirs are also considered an important source of methane (Rudd et al. 1993). Of the major greenhouse gases, the concentrations of methane can be stabilized relatively easily by modest cuts in anthropogenic emissions. A 10–20 per cent cut would suffice to stabilize concentrations at current levels due to methane's relatively short atmospheric life-time, assuming that this life-time remains constant and that the natural emissions do not change. Whether this does actually happen will depend on how the tropospheric chemistry is influenced by emissions of hydrocarbons and carbon monoxide among others and on whether change of global climate itself affects methane emissions.

Nitrous oxide (N_2O): The concentration of nitrous oxide has increased by 5–10 per cent since preindustrial times. The exact cause cannot be pinpointed but the use of nitrogenous fertilizers, land-clearing, biomass-burning and fossil fuel combustion may all have contributed. Each additional molecule of N_2O has over 200 times as much impact on climate as an additional molecule of CO_2. Nitrous oxide also contributes to the stratospheric ozone depletion. N_2O is currently increasing at a rate of 0.25 per cent per year, which represents an imbalance of about 30 per cent between total emissions and total losses. Table 1.5 represents the global N_2O emission estimates. N_2O increases are responsible for roughly 6 per cent of the current increase in the commitment to global warming. Assuming that the observed increase in N_2O concentration is due to anthropogenic sources and that natural emissions have not changed, then an 80–5 per cent cut in anthropogenic emissions would be required to stabilize N_2O at current levels.

Halocarbons: Chlorofluorocarbons (CFCs) are well-known as the depleters of stratospheric ozone but they are also significant as greenhouse chemicals. The most common species of CFCs are

Table 1.5 Global N_2O emission estimates

Sources/sinks	Range $(10^6 ta^{-1})$*
Sources	
Oceans	1.4–2.6
Soils (tropical forest)	2.2–3.7
Soils (temperate forests)	0.7–1.5
Fossil fuel combustion	0.1–0.3
Biomass burning	0.02–0.2
Fertilizer (including ground water)	0.01–2.2
Sinks	
Removal by soils	Unknown
Photolysis in the stratosphere	7–13
Atmospheric increase	3–4.5

Source: Houghton et al. 1990
*ta^{-1}: tonnes per annum

CFC-12 (CCl_2F_2) and CFC-11 (CCl_3F). Their atmospheric concentrations in 1986 were 392 and 266 parts per trillion by volume. While these concentrations are tiny when compared with that of CO_2, CFCs have as much as 20,000 times more impact on climate per additional molecule than CO_2. Further the atmospheric concentration of CFCs is increasing very rapidly—more than four per cent per year since 1978—representing about 15 per cent of the current increase in the commitment to global warming. It is estimated that 75 per cent and 85 per cent cuts respectively, are required in the emissions of CFC-11 and CFC-12 to stabilize the atmospheric concentrations to their present levels. However, because of contributions from several other compounds, in order to stabilize the total greenhouse warming potential from all halocarbons, a phaseout of fully halogenated compounds (those that do not contain hydrogen), a freeze on the use of methyl chloroform and a limit on the emissions of partially halogenated substitutes would be required.

Other gases influencing composition: Increasing emissions of carbon monoxide (CO) and nitrogen oxides (NO_x) are also adversely influencing the chemistry of the atmosphere. This change in atmospheric chemistry alters the distribution of ozone and the oxidizing power of the atmosphere, changing the atmospheric lifetime of the greenhouse gases. If the concentrations of the long-lived gases were stabilized, it might only be necessary to freeze emissions of the short-lived gases at current levels to stabilize atmospheric composition.

Many of the greenhouse gases that induce tropospheric warming through the greenhouse effect are highly stable and may be retained in the atmosphere for decades or even for a century or more. Though the role of CO_2 in global warming appears substantial at present due to the high concentrations of the gas, gram to gram, other greenhouse gases (GHGs) are a thousand to ten thousand times more effective than CO_2 and are consequently dangerous even at their present trace levels of concentration.

The concept of relative GWPs (Global Warming Potentials) has been developed to evaluate the relative radiative effect (and, hence, the potential climate effect) of equal emissions of each of the GHGs. The GWPs take into account the differing residence time of gases in the atmosphere and define the time-integrated warming effect due to an instantaneous release of a unit mass (1 kg) of a given GHG in today's atmosphere, relative to that of CO_2. The relative cumulative effect of man-made emissions on the climate is summarized in Table 1.6.

GLOBAL WARMING AND THE GREENHOUSE EFFECT

There are a number of questions concerning these phenomena still puzzling the scientific community.

Where is the evidence of global warming?

Shouldn't our atmosphere's ability to trap heat, the sign of the greenhouse effect, be noticeable by now? The 1980s saw the four warmest years recorded in the last 110 years. The first seven months of 1988 were hot enough to break all records and caused some to declare that the greenhouse effect was now being felt. Unfortunately the problem of distinguishing a long term warming trend caused by increasing CO_2 from the natural variations in the earth's climate is a formidable one. Serious droughts and heat spells have occurred from time to time in the past. How can one be sure that the weather in the 1980s reflected warming due to human alteration of the atmosphere, especially when 1989 was relatively cooler? Nevertheless, some researchers estimate that a warming of 1°F has already occurred during the last century. But because of the difficulty of comparing past and present measurements, this finding is far from being universally accepted.

Table 1.6 The relative cumulative climate effect (1990) of man-made emissions

	GWP (100 yr horizon)	1990 emissions (Tg)	Relative contribution over 100 years
Carbon dioxide	1	26000	61%
Methane*	21	300	15%
Nitrous oxide	290	6	4%
CFCs	varies	0.9	11%
HCFC-22	1500	0.1	0.5%
Others*	varies		8.5%

GWP = Global Warming Potential
* These values include the indirect effect of these emissions on other greenhouse gases via chemical reactions in the atmosphere. Such estimates are highly model-dependent and should be considered preliminary and subject to change. The estimated effect of ozone is included under 'Others'
Source: Houghton et al. 1990

CLIMATIC MODELS AND THEIR FORECASTS

Virtually all scientists agree that if 'greenhouse gases' increase and all other factors remain the same, the earth will warm up. The ways in which climatic forces interact are poorly understood and it is possible that some factors may counteract warming just as some other factors may enhance it.

Because the warming may be hard to detect with certainty until well into the next century, climatologists have turned to computer models to make rough predictions of how much warming will occur and where and when it should become noticeable. A major shortcoming of the computer models however has been that they have focussed almost entirely on the atmosphere's response to the rising levels of CO_2, neglecting the oceans. Although slower to react to change than our gaseous atmosphere, the oceans are nonetheless active components of the globe's climatic system and can influence long-term temperature variations. The oceans' most important effect in this case is their ability to absorb, release and redistribute heat. Thus, the response of the oceans to increased greenhouse gases is a critical unknown in most climate models. The reason for this is that the forces at work in the oceans are not nearly as well observed and understood as are those of the atmosphere. Modelling such coupled ocean-atmosphere systems simultaneously is a daunting task.

Energy balance (EB), radiative-convective (RC) and general-circulation models (GCMs) have been developed to simulate conditions of global warming (Budyko 1969; Sellers 1969). In 1975, Syukuro Manabe and Richard Wetherald of the National Oceanic and Atmosphere Administration's Geophysical Fluid dynamics lab were the first to create a three dimensional climate model to study the greenhouse effect. The ocean component in their model was a swamp—a stagnant, wet surface that absorbed and released heat and moisture. The swamp model calculated the ocean's surface temperature by balancing such factors as solar radiation, infra-red radiation to and from the earth, evaporation and heat transfer between sea and air. Other important ocean processes such as currents and salinity, seasonal changes and stored heat were not factored at all.

The late 1970s saw the development of models that treated the upper ocean as though its layers were mixed together. These models assumed an ocean that was a slab 150 ft deep. This approach allowed for heat storage in the summer, and release to the atmosphere in winter, thus reproducing seasonal features at least.

The basic physics of ocean circulation first explained by Count Rusuford (Benijanin Thompson) in 1800 is only now being incorporated into coupled atmosphere-ocean climate models. The polar to equatorial currents and ocean circulation, where cold, highly saline water sinks and warm fresh water floats on the surface, is a complex way in which local atmospheric conditions help to determine the temperature and salinity of an ocean's surface waters. Thus, the rapidly changing atmosphere works to influence the formation of deep ocean currents that may take tens of thousands of years to circulate. In 1984, James Hansen (NASA Goddard Institute of Space Studies, New York) developed an atmosphere–ocean climate model. Even though his model did not explicitly calculate ocean currents, it took into account the heat they transfer from the tropics to the poles. This model again had limited usefulness, because in reality the heat transfer would vary as the ocean currents adjust to the altered atmospheric conditions.

Thus, only recently have climatologists attempted to run atmospheric models simultaneously with dynamic models of the ocean. These models can simulate temperature and salinity-driven deep ocean currents as well as allow for upper layer wind-driven surface currents (such as those created by Westerlies). One such model was developed and run at the National Center for Atmospheric Research in Boulder, Colorado. Starting off with present levels of CO_2 and at a 1 per cent rate of increase for the next 30 years, they found that the lower atmosphere and the ocean surface gradually warmed up to a global average of 1°F (Washington 1992). But the warming was not equally distributed —some regions, mainly the continental interiors, became as much as 7°F warmer. Towards the end of the computer runs, however, a seemingly paradoxical cooling appeared in Northern Europe where temperatures dropped by as much as 11°F during winter months. These results are not seen in computer simulations that use simple and mixed/stratified layer ocean models. That this cooling appears only in the model that incorporates the effects of the ocean currents is reason to suspect that the ocean circulation is lessened by increased CO_2 concentration. But figuring out exactly which elements of our climate simulation were responsible, would require examining how ocean currents affect the temperatures on land in Northern Europe and then, how these currents might have been altered by a generally warming climate.

Even the best among such models still oversimplify ocean-circulation, precipitation, moisture-exchange near the atmosphere-ocean-land interface, and the role of clouds. At any given time for example, clouds cover 60 per cent of the planet not only trapping heat radiating from its surface, but also reflecting sunlight back into space. So, if the oceans heat up and produce more clouds through evaporation, the increased cover might act as a natural thermostat and keep the planet from heating up. After factoring more detailed cloud simulations into its computer models, the British Meteorological Office recently showed that current global warming projections could be cut in half.

Oceans have a major effect upon climate, but scientists have only begun to understand how. Investigators at the US National Center for Atmospheric Research attributed the North American drought in the summer of 1988 primarily to temperature changes in the tropical Pacific involving a current called *El Nino*—not to the greenhouse effect. As the oceans shuffle global climates into new patterns, the warming produced by the greenhouse effect may be even harder to detect and some areas may actually cool. Some have suggested that if the earth had no oceans at all we would have been able to measure the greenhouse warming with certainty by now. Oceans not only have the ability to alter climatic patterns, but they can also slow down the warming effects, by absorbing a great deal of heat. Because the Southern Hemisphere contains substantially more amount of ocean than the Northern Hemisphere, it probably will not warm as rapidly. This may seem like good news, but eventually the heat stored in the oceans will affect our climate, perhaps not in a few years but decades and centuries from now, given the time scale of ocean circulation.

Is global warming a figment of the imagination?

Two widely reported statistics *seem* to present a powerful case for global warming. Some temperature records show about half-a-degree of warming over the past century, a period that has also seen a noticeable increase in greenhouse gases. Six of the warmest years globally, since record keeping began 150 years ago, have all been in the 1980s.

US climate experts report that warm temperatures dominated the global climate in 1994—similar to the record levels that were observed in 1990 and 1991. David Rodenhuis,

Director of the Climate Analysis Center at the National Weather Service told reporters that after two relatively cooler years the estimated mean global temperature during 1994 was 0.4°C above normal. Global temperatures from March to December were the warmest observed since 1951.

Rodenhuis said that the rebound of above-normal temperatures in 1993 may have been partially due to the fact that the shading effects of the aerosols from the 1991 eruption of Mount Pinatubo have ended. Global temperatures achieved a maximum again in 1995, it being the warmest year globally since records were first kept in 1856, according to a provisional report issued by the British Meteorological Office and the University of East Anglia. The average temperature was 58.7°F, according to the British data. The figures also reveal the years 1991 through 1995 to be warmer than any similar five year period, including two half-decades of the 1980s, the warmest decade in the record to date.

According to a recent report (*Down to Earth* September 1997), the climate is getting warmer and the effects of global warming are becoming increasingly noticeable in countries such as UK which are known to have a cold climate. Scorpions, natives of the warmer climates, are thriving in the Southern county of Essex. Frogs and toads have been laying eggs earlier than usual. Birds are also laying eggs much earlier than in the past.

Yet there is a school of thought—the subscribers to which are not many—that the greenhouse effect may well be a grossly exaggerated fear if not entirely a figment of the imagination. This school of thought cites older records of temperature change which do not seem to fit in with the greenhouse theory. Between 1880 and 1940, temperatures appeared to rise. Yet between 1940 and 1965—a period of much heavier fossil fuel use and deforestation—temperatures dropped, which seems inconsistent with the greenhouse effect. And a comprehensive study of past global ocean records by researchers from Britain and the Massachusetts Institute of Technology (MIT) revealed no significant rising-temperature trends between 1856 and 1986. Taking everything into account not all climatologists are willing to attribute the seeming warming to the greenhouse effect. In May 1989, 61 scientists participating in a greenhouse workshop in Massachusetts declared that 'such an attribution cannot now be made with any degree of confidence'. However, the relationship between the emission fluxes of greenhouse gases and the actual increase in atmospheric concentration depends, in a very complex manner, on a number of factors in which atmospheric chemistry plays a very important role (Wuebbles et al. 1989).

Some scientists consider the process of global warming as yet another phase in the life of this planet, a simple cyclical process, which would ultimately revert to its original state. And Paul Colinvaux has even said 'those self-styled spokesmen for the ecological profession who warn people that the industrial life threatens to destroy the atmosphere, spread the gospel of nonsense. It would be better for us if they ponder the fable of the little boy who cried "wolf, wolf".'

This brings one to the provocative conclusion that, although air is regulated in the very long term by the doings of living things, and might even have been made by ancient ecosystems in the first place, the accumulated store of gas is now so vast as to be almost independent of life processes for its maintenance. As seen earlier, the tiny concentration (0.03%) of carbon dioxide is maintained by an elaborate way of interacting mechanisms, rather as the nutrient supply of the oceans is kept dilute and constant over geologic time. The amount of carbon dioxide in solution in the oceans will be equivalent to five times the total weight of the atmospheric gases. The gas is in equilibria with the atmosphere and the oceans. If the atmospheric CO_2 drops, the sea loses CO_2. The chemistry of its solution equilibria is not simple, since it involves carbonic acid, carbonates and bicarbonates. If the atmospheric CO_2 rises, the ocean soaks up the excess. Thus, the chemical equilibria between the air and the sea should be very hard to imbalance. Nor is this all, because the shock-absorber is itself shock-absorbed. Looking back at the figure on the carbon cycle, we see that the CO_2 in the oceans is itself kept constant by formation of carbonate that settles on the sea bed. Thus, the regulation of CO_2 the air is largely determined by chemical and physical processes above the surface of the earth. All life adapts to what these processes give it.

Yet there is one caveat, and it concerns CO_2 supply. We now use petroleum and coal for fuel. When we burn it we add fresh CO_2 to the air. This CO_2 was in the air in the first place, of course millions of years ago, but the plants that produced sugar from it died in places where they could not rot and so were converted into coal or oil. At the same time this loss to the air was made good by the ocean shock absorber and the morsel was 'forgotten' by the system. But now when we burn it, that ancient CO_2 comes as something new, a fresh influx of gas to the air. Again our ocean shock absorber shall come to the rescue, soaking up this excess CO_2. But some recent measurements of CO_2 in the air by geochemists have been surprising all the same. The atmospheric CO_2 has been steadily rising from year to year. We had placed our faith in the shock absorbers a little too soon.

What has gone wrong with the absorbers then? Those scientists who believe that global warming is just a passing phase, say that it takes time to work. The oceans have to be stirred up so that water with the extra dissolved CO_2 can get to the bottom and be replaced by other water to take its turn in the blotting paper queue. The oceans which are 5 miles deep and stirred by surface winds take centuries to stir up. The shock absorber is working, but slowly. In the end it will soak up all the polluting CO_2 from our fossil fuels and pass it to its mud, but meanwhile the shock absorber has more than it can handle, thus, before the absorber can catch up, maybe the CO_2 levels will have doubled to about 0.06 per cent. This will not be a catastrophe for life. On the other hand it will certainly make a difference. The worry is that we do not know what all the effects might be.

LIKELY IMPACT OF A 'WARMED' EARTH

The GCMs (general circulation models) on which predictions of climate change are based, unfortunately, have not attained a high degree of reliability with respect to regional climate effects but there is a general convergence in their main findings which are briefly summarized below.

Changes in climate: The effects of a warmer atmosphere on ocean currents, with the net result of causing distinct climatic patterns the globe over have already been discussed in detail. The strengthening of the Indian monsoons due to a warmer regime is also one of the proposed effects (Ramanathan 1985); the results of which may manifest as a greener country more like a tropical evergreen rain forest. However, as mentioned earlier, higher temperature, coupled with increased precipitation will lead to a characteristic shift from C_4 to C_3 plants (Sukumar 1993). The C_3 plants (wheat, rice, legumes, oil seeds, cotton) show higher CO_2 assimilation, growth and yield in response to higher CO_2 concentration, than the C_4 crops (sugarcane, sorghum or maize).

Warning against compelling evidences of a shift in the Earth's weather patterns and changes in climate, the World Wide Fund for Nature recently released the 'State of the Climate Report' (1997), compiling a huge array of global data which clearly signals that a change is already underway. Stating that climate change was affecting every region and most nations (*The Hindu*, 8 Oct. 1997), the report said that 1995 was the hottest year in history and 1997 looks set to be a close second. The five hottest years have all occurred in the 1990s. Excerpts from the report are presented below.

The world is experiencing the biggest thaw since the last Ice Age. Much of Siberia is three to five degrees celcius warmer than it was earlier this century. Europe's alpine glaciers have lost half their volume since 1850, glaciers in the Peruvian Andes are retreating, and the US government predicts there will be no glaciers left in the Glacier National Park in Montana by 2030. In Antarctica, some penguin populations have crashed and krill populations—a food-source for many marine animals—have declined, seemingly killed by warmer waters.

Much of the tropics have become hotter and drier—especially in the already arid region stretching from West Africa to Indonesia. In the 1990s, southern Africa suffered crop failures, water shortages and the five driest years of this century. The Gobi region of Mongolia has been steadily getting less summer rainfall in the last 30 years. And the drying trend extends into Europe. Rainfall is down by 20 per cent in the Mediterranean regions. Spain had five years of continuous drought beginning in 1991. In Greece, the flow of the nation's longest river, the Acheloos, declined by 40 per cent in four years.

The report also claims that the catastrophic fires currently (1997) raging in Indonesia with a crippling effect on people's health is a result of changing climate patterns, particularly global warming.

Changes in sea level: Another effect that is expected to occur is the rise in the sea-levels, due to the melting of the polar ice-caps. Some people have started to claim its action rightaway; but how much of this can be attributed to global warming is not known. A number of models have been developed to compute the magnitude of snow-melt causing the rise in sea-level. Revelle (1983) derived a sea-level rise of about 70 cm for a global warming of 6°K. A full scale study done in Denmark went on to prove little, ending with the conclusion that Denmark as such would be only slightly affected by an average increase in global temperatures (Fenger et al. 1993).

A rise in sea level represents a potential threat to existing coastal, economic, social and environmental systems. According to a scenario developed by IPCC (Inter Governmental Panel on Climate Change), global warming is predicted to cause an increase of global mean sea-level by 65 cm, by the year 2100 (IPCC 1990). The primary effect of rising sea-level will be increased coastal flooding, erosion, storm surges and wave activity. These primary effects will lead to loss of ecosystems such as wetlands, loss of coastal vegetation and habitats, salt-water intrusion into groundwater systems, and the loss of cultivable land. Such changes would, in turn, translate into socio-economic effects. This prospect assumes alarming significance in the context of the South Asian countries, particularly with respect to island nations like Maldives. Such islands which are barely two to four metres above sea level, could become extinct if the sea level rises as per the forecasts. The entire coastline of South Asian countries, which is thickly populated, would also face serious disturbances. Sea level rise is likely to increase the vulnerability of this region to tropical storms, storm surges and greater inundation. Some of the mangrove forests of this region are likely to be completely decimated (Pachauri 1992).

Impact on forests: The Amazonian interior uses 80 per cent of incident radiation for evapotranspiration, the rest 20 per cent being used to warm the air. If deforested, the reduction in precipitation will lead to release of latent heat. Thus, the regions outside the tropics will receive less heat and become cooler.

Impact on agriculture: The global warming will inevitably change the climate and affect food production, particularly in the tropical countries of Africa, South America and Asia, where the productivity is already poor (Parry 1990). Contrary to this, a uniform or general rise in global temperatures may shift the vegetation belts to higher latitudes, especially shifting of the temperate wheat zones further north. Poorer soils can reduce productivity. Yet evidence also exists that increased CO_2 accelerates growth of trees. Pine trees that have barely survived on the timberline of California's mountains are growing luxuriantly now (Friedmann 1985). Since rapid plant growth can remove CO_2 faster from the atmosphere, it can slow the climatic impact of the greenhouse effect. Agricultural scientists generally take a favourable view of the increase in CO_2. Not only will more CO_2 make crops grow faster and larger but they will also become more drought resistant because higher CO_2 concentrations also improve the efficiency with which plants use water.

Impact on ecosystems: Effects on animal life are more or less ambiguous. Sex ratio biases in certain reptiles like turtles could occur. However, more detailed research needs to be put in before any conclusions can be drawn.

POLICIES NEEDED

Being so unsure about the effects of global warming as well as global warming itself, it has become very difficult to take any preventive measures. But discretion is the better part of valour. Discouraging the use of coal and other fossil fuels and developing alternate forms of energy may prove beneficial.

Many countries have levied CO_2 taxes. Norway charges 120 $/tonne of C, Sweden charges 150 $/tonne C; yet if we have to stabilize global CO_2 levels, a global average of 210 $ has to be charged (USEPA 1989). Biomass conservation and farming can sequester a good deal of CO_2 (Rosillo Calle et al. 1992). We also have to stop deforestation, at least lower the rates of deforestation of tropical rain forests and grow or afforest at least

30 million ha in developing nations and 40 million ha in industrial nations. Of course the major conventions, treaties, and protocols continue side by side. Many of the policies could be deleterious to the furthering of development in many Third World Countries. In fact there is the major danger of the developed nations making use of the 'wolf, wolf' strategy to achieve a new world order of Environmental Imperialism.

The Montreal protocol signed by 40 countries in September, 1987, embodies the strategy to eventually phase out chlorofluorocarbons (CFCs) by the year 2000. In the Indian context, the Montreal protocol seeks to phase out production of CFCs in the developing countries by the year 2010 (a grace period of 10 years). The cost of switching over to CFC substitutes is estimated to be between Rs 35,000 million and Rs 60,000 million depending on the time-lag for the change-over. India cannot have access to the technology to produce the replacement refrigerants as it has been patented by Dupont and a few other multinational corporations.

The agreement adopted at the UN climate conference (Berlin, April 1995) includes a tool to give poorer countries their best access to rich nation's technologies. It is a mechanism set up so that clean energy technologies can be transferred to the developing world, with benefits for both parties and the global environment. Nearly 120 nations adopted a mandate at the Berlin meet on strengthening the 1992 Rio Climate Treaty. It obligates developed countries to set 'objectives' for further emission cuts by 1997. The Third Conference of the parties to the United Nations Framework Convention on Climate Change was recently held (December 1997) at Kyoto, Japan. As a result of the deliberations, the USA, the European Union, and Japan have committed themselves to cuts of seven, eight and six per cent of the 1990 levels of six greenhouse gases over the next 15 years: hydrofluorocarbons, perfluorocarbons, sulphur hexafluoride, carbon dioxide, nitrous oxide and methane. It may be noted that prior to the Kyoto convention only the last three gases were identified for emission control. Another aspect of the Kyoto protocol is that the right to emissions trading was included in the final treaty. Perhaps the summit's greatest achievement was its eloquent reiteration that global warming is not just a figment of a science-fiction writer's imagination, but a painfully real problem.

On the scientific side, tremendous research with remarkable achievements in alternative energy sources; less polluting, more efficient systems, is going on. Nothing much is being done to actually reduce atmospheric CO_2 levels. Sequestering CO_2 in new forest growth can offset anthropogenic emissions by fixing carbon in plant tissue. Long-term sequestration requires that forests are periodically harvested for lumber and wood products that remain in service and do not return CO_2 to the atmosphere by combustion. Simple prudence and being more efficient can alter matters to an extent, but the rising trend of CO_2 emissions is something about which not much can be done now. Global warming responses should be linked to existing programmes for energy conservation, soil and water conservation, sustainable agriculture and sustainable urban development. Some remedial measures based on the 'Report of the 1991 Woodlands Conference' (Jurgen et al. 1992) are presented below which address agriculture and forestry, sea-level rise, population and nutrition, energy and research needs.

Agriculture and forestry: Maintaining and increasing the diversity of ecosystems, crop varieties, technologies, and farming systems are keys to reducing vulnerability to global warming. Policies should emphasize the production of biodiversity through species and ecosystem conservation programmes. Special conservation programmes for habitats in which the microclimate is cooler than the regional macroclimate are needed. Crop genetic diversity and respect for traditional agricultural knowledge and technology can be enhanced by developing flexible agricultural systems and options for new food supplies. Each region should monitor its forests, streams and soils for early drought warning signs. Should drought occur, policy measures to reduce water use should be implemented.

Changes in forests and agricultural systems must be linked when assessing the impacts of climate change. These linkages can be studied by examining the impacts of climate change on agro-forestry. Agroforestry can utilize a traditional system that combines aspects of agriculture with forest management procedures. To overcome the barriers to natural ecosystem migration, policies should be aimed at introducing new species into these systems that can expand buffer zones in preserved or managed areas. Transplanting or assisting in the migration of species might be useful in commercial fishing and agriculture, and in saving some endangered species.

Rising sea level: Improved regional storm prediction is needed, especially in regions that are particularly vulnerabale to typhoons, hurricanes, tropical storms and other severe storms. One example is the reported increase in the number of tropical storms to hit the China coastline over the past century. Co-operative research should be strengthened on the many areas that share both biophysical characteristics and vulnerability to the adverse effects of rising sea levels. This should include closer co-operation among international agencies and regional assessment entities.

Strategies to limit future development in vulnerable coastal environments must be given a high priority. Consideration should be given to locating new, major infrastructure inland from vulnerable areas. This could be enhanced by removing government incentives like insurance subsidies that allow development and redevelopment in vulnerable areas. Accelerated sea-level rise will aggravate problems associated with existing development.

Integrated coastal management zones are essential. These should include not only coastal fringe areas, but also those inland areas that affect the coastal fringe. Land-use strategies should protect natural ecosystems and avoid, where possible, non-sustainable, rigid solutions. They should include measures to minimize pollution and storm risks. Once designed, coastal land-use management strategies should be executed immediately. In the UK, preparations are on to counter the rise in sea level. New regulations require that all new sea defences being built in eastern England to be about 25 cm higher than earlier. Another interesting measure being adopted is known as the 'coastal retreat'. Under this, sea walls are being breached to let the sea back into reclaimed areas in East Anglia. New marshes are being created to serve as a soft buffer from the sea. The marshes take the energy out of the waves before they attack the new sea walls situated further inland. The buffer also serves as a habitat for wild life (*Down to Earth* 1997).

Pollution, nutrition and health: Both excessive population growth and increases in per capita consumption levels contribute to the growth in greenhouse gases, and increase the vulnerability of regions and societies to global warming. In some regions, population growth, immigration and the conditions of land and resource distribution mean that pressures exerted by local populations are approaching or exceeding the carrying capacity of land and water. Where global warming is likely to further degrade a strained resource base, excessive population growth could bring regional crises and migration to other regions. Some of these relocations could be to ecologically vulnerable regions, such as the Amazonian interior, the Indonesian coastal regions, and other coasts, forests and marginal agricultural and grazing lands. To enhance sustainable development it is important that policies to reduce rapid population growth be encouraged and integrated with other programmes that reduce vulnerability and excessive per capita consumption of energy and resources.

Energy: Society's accelerating use of fossil fuel energy is the principal contributor to greenhouse gases. In industrialized regions, the extraordinarily high level of energy consumption is not sustainable and creates additional environmental problems in addition to contributing to climate change. Significant energy-efficiency gains are possible, and major opportunities exist to develop both new energy-efficient supply and end-use technologies. The development of renewable and other alternative energy supply technologies also needs to attract greater support from government and industry. Industrialized, economically affluent nations have a special responsibility to carry out the necessary research and development.

Research and education: There is a shocking lack of basic data on natural ecosystems at all levels. Programmes to obtain statistically valid, temporally consistent data-monitoring for forests and other natural ecosystems are essential in developing realistic policies for managing these resources. To better understand complex ecological systems and their interactions with the climate, we need an international network of regional-global environment research centres. We also need improved or new regional climate monitoring by networks dedicated to global and regional climate change studies and their impacts. Scientists should be encouraged to develop finer resolution and mesoscale models, shed light on weather patterns, storm frequencies, and drought persistence at the regional level. There is also a need to move beyond the static doubled CO_2 model to recognize the dynamic nature of the buildup and the likelihood that we will exceed a doubling of CO_2.

The fact that remedial measures are being contemplated globally, without waiting for the nexus between greenhouse effect and global warming to be proved unambiguously, augurs well for the future.

REFERENCES

Bolin, B. (1970) 'The Carbon Cycle'. In *The Biosphere*. New York: Scientific American Publishers.

Colinvaux, P. (1989) *The Regulation of Air*. Princeton, New Jersey: Princeton University Press.

De, A.K. (1986) *Environmental Chemistry*. New Delhi: Wiley Eastern Publishers.

EPA (1989) *'Global Warming and Policy Decisions'*. Report of the US EPA. Washington D.C.: EPA.

Fenger, J., A.M.K. Jorgensen, H.E. Mikkelsen and M. Philipp (1993) 'Greenhouse Effect and Climate Change—Implications for Denmark'. *Ambio* 22.

Friedman, H. (1985) 'The Science of Global Change—An Overview'. In *Global Change*, edited by T.F. Malone and J.G. Roederer. Cambridge: Cambridge University Press.

Golitsyn, G.S. (1993) 'The Changing Atmosphere'. In *Global Change*, edited by Malone, T.F. and J.G. Roederer. Cambridge: Cambridge University Press.

Hilbertz, W.H. (1992) 'Solar Generated Building Material from Seawater as a Sink for Carbon'. *Ambio* 21(2).

Houghton, J.T., G.J. Jenkins and J.J. Ephraums (eds) (1990) *Climate Change*, The IPCC Scientific Assessment Report. New York: Cambridge University Press.

Jorgensen, S.E. (1983) 'Modeling the Effect of CO_2 Pollution on Climate'. In *Application of Ecological Modelling in Environment Management*. Amsterdam: Elsevier Scientific Publishing Company.

Oort, A.H. (1970) 'The Energy Cycle of the Earth'. In *The Biosphere*. New York: Scientific American Publishers.

Parry, M.L., J.H. Porter and T.R. Carter (1990) 'Climatic Change and its Implications for Agriculture'. *Outlook on Agriculture* 19.

Rossillo, F. Calle and D.O. Hall (1992) 'Biomass Energy, Forests and Global Warming'. *Energy Policy* 20.

Rofmans, J., M.G.J. den Elzen, M.S. Krol, R.J. Swart and H.J. Van der Woerd (1992) 'Stabilizing Atmospheric Concentrations: Towards International Methane Control'. *Ambio* 21.

Rudd, J.W.M., R. Harris, C.A. Kelly and R.E. Hecky (1993) 'Are Hydroelectric Reservoirs Significant Sources of Greenhouse Gases'. *Ambio* 22.

Sedjo, R.A. (1992) 'Temperate Forest Ecosystems in the Global Carbon Cycle'. *Ambio* 21.

Sukumar, R., R. Ramesh, R.K. Pant and G. Rajagopalan (1993) 'A $\delta^{13}C$ Record of Late Quaternary Climate Change from Tropical Peats in Southern India'. *Nature* 364.

Washington, W.M. (1990) 'Where is the Heat'. *Natural History* 3.

Wuebbles, D.J., K.E. Grant, P.S. Connel and J.E. Penner (1989) 'The Role of Atmospheric Chemistry in Climate Change'. *The Journal of the Air and Waste Management Association* 39.

2

Acid Rain

Pollution is like a boomerang. We fling it away from us with all the force at our disposal...only to have it come back and hit us at unexpected spots and in an unpredictable manner. Like the nemesis.

Acid rain is an example of pollution we send away...far away. And it comes back to haunt us. *Acid rain* is normal rain acidified by certain air pollutants. When it falls to the earth it dirties and damages the environment instead of cleaning and enriching it.

What is acid rain?

The measure of acidic strength of any medium is done through its pH (the negative logarithm of the hydrogen ion activity of the medium). If the pH is more than 7 the medium is alkaline (or basic). A pH of less than 7 indicates acidic character. The lower the pH, the stronger the acidity. A substance with a pH of 1 is very strongly acidic. A substance with a pH of 6 is only mildly or 'weakly' acidic.

Rain occurs when water vapour condenses in clouds and falls to earth. As it begins to fall the rain is neutral—neither acid nor alkaline. While it travels through the air, it dissolves floating chemicals and washes down particles that are suspended in the air. In clean air the rain picks up only materials that occur naturally, such as dust, pollen, some carbon dioxide (which forms the mild carbonic acid) and chemicals produced by lightning and volcanic activities. These substances make the rain slightly acidic, with a pH of about 6. This level of acidity is not considered dangerous. *Acid rain* is rain with a pH of less than 5.6.

How does acid rain form?

When rain falls through *polluted* air, it comes across more acid-forming substances and in higher concentrations than otherwise. Among the chemicals frequently occurring in polluted air at higher-than-normal concentrations are sulphur oxides and nitrogen oxides. In some situations hydrochloric acid vapour and mists of other acids such as phosphoric acid may also be present. These gases dissolve in falling rain making it more acidic than natural rain. This leads to *acid rain*. Acid fog is formed when chemical pollutants are dissolved in very moist air. This causes changes in the pH of the air in the same way that acid rain changes the pH of the soil or the water in lakes and rivers.

When fossil fuels—oil, coal and gas—are burnt in industrial plants or automobile engines, large amounts of sulphur oxides and nitrogen oxides are released into the air. A single industrial smokestack can produce as much as 500 metric tonnes of sulphur oxides each day. Most of it is sulphur dioxide, which becomes sulphuric acid in moist air.

Nitrogen oxides and carbon particles produced by automobiles and other motor vehicles are so plentiful in some large cities that the air is coloured by them. Nitrogen oxides also enter the air through the breakdown of agricultural fertilizers. These eventually form nitric acid in moist air. Like sulphuric acid, nitric acid corrodes many metals, and strong concentrations of nitric acid can cause burns on skin.

Amongst natural sources of acid rain, other than volcanoes and geyzers which also contribute SO_x and NO_x (sulphur and nitrogen oxides), the most significant is formic acid (Sanhueza 1991). Biomass burning due to forest fires causes emission of formic acid (HCOOH) and formaldehyde (HCHO) into the atmosphere. A large fraction of formaldehyde gets oxidized to formic acid *in the atmosphere*. During the days prior to the rainy season in tropical areas, forest fires are common. In this period photochemical activity in the atmosphere is also high. Thus, higher emissions of HCOOH and HCHO, combined with swift conversion of HCHO to HCOOH due to photo-oxidation, causes acidity in the rain during the first few days of rainfall over tropical forests.

Millions of metric tonnes of sulphuric acid and nitric acid fall in rain on earth each year, principally in the United States, Canada and Europe (Moiseenko 1994). The concentrations of these acids are too weak to cause burns, but they do produce other severe effects. Other acids also occur but in smaller amounts. Hydrochloric acid is often produced directly by smokestacks. Carbon monoxide and carbon dioxide are produced by automobiles. These become carbonic acid. The acids generated in the air due to such pollution have changed the normal pH of rain from 5.6 to an average of 4.5 over the entire eastern United States and Canada. A pH of 4.5 is more than ten times as acidic as a pH of 5.6. Some areas have received acid rain with a pH of 3.0, which is as acidic as vinegar.

One of the areas thought to be most severely affected by acid rain is the scenic Adirondack mountain region of New York State. But there are no heavy industries or factories in this area and not enough automobiles to produce much pollution. Where does the acid rain in the Adirondacks come from? The answer is that pollutants can travel over hundreds and even thousands of kilometers. Many factories have smokestacks well over 100 metres (328 feet) high. Pollutants in smoke enter the air high above the ground and are carried by winds to great distances.

Pollution from Canada is carried by winds into the United States. Norway and Sweden have received acid rain created by pollutants travelling north from England, Germany, Italy, France and Austria. These happenings have made the world more aware than ever before, of the global impacts of local environmental problems and have inspired nations across the world to take joint initiatives in mitigating the damages caused by acid rain.

A BRIEF HISTORY OF ACID RAIN

The earliest documented use of the term *acid rain* is found in a 1872 book *Air and Rain: Beginnings of Chemical Climatology* authored by R.A. Smith (Soni 1991, Hidey 1995). Even prior to that Smith had been studying atmospheric acidity in the industrial town of Manchester, England, for several years and had suspected that the soot of the industries was responsible for the rain-water acidity. He found that sulphuric acid was present in the rain-water falling over Manchester (Hidey 1995).

During the 1950s, investigations on mysterious fish deaths, suspected to be associated with water acidification in Norway and Sweden, led Scandinavian scientists to express concerns about the origins of the acidity. Around the same time Gorham (1955) documented acidification of rain falling over England. Such efforts were stimulated in the 1960s by the more obvious links seen between fish (especially salmon and trout) deaths in Scandinavian lakes and acid deposition. The early 1960s also saw the symptoms of acidification of rain appearing in USA. With every passing year more and more evidence of acid rain piled up. The concerns grew in intensity to a point when the issue of acid rain was raised in the 1972 United Nations Conference on the Environment. The reason for the international debate was the belief that the transport of air pollutants from some countries was acidifying the waters of some other countries.

By the mid-1970s, surveyors in northeast America began to report observations of apparent acidification of remote lakes and the parallel deterioration of fisheries. The deterioration in the water quality of lakes in southern Ontario and the Adirondack mountains caused particular concern. A few years later, forest scientists in Europe and North America began to speculate about the role of acid deposition in forest dieback. As a result of international concerns, aggressive research programmes to characterize the extent and severity of acid rain exposure were initiated in both Europe and North America. Mounting circumstantial evidence that pollution from industrialized areas could effect remote regions far distant from the points of origin of the pollutants prompted efforts towards internationalisation of control strategies for sulphur and nitrogen oxide emissions.

In the 1980s, the United States of America initiated one of the largest publicly funded research programmes ever attempted on a single issue—acid rain. After a 10 year Congressional authorization, US National Acid Precipitation Programme (NAPAP) completed its work. This effort was closely paralleled by a Canadian research program. These two projects provided a large body of knowledge about acid deposition and its environmental effects in much of North America. NAPAP was reauthorized to investigate and report on the effectiveness of the acid rain reduction programme legislated in 1990 for the United States.

From the mid-1980s developing countries, such as India, began monitoring rain-water quality with particular emphasis on acidification. These efforts have grown in intensity over the years and so have efforts to control acid-causing pollutants.

Acid rain in the twentieth century is essentially a product of aggressive industrialization in the developed countries. It is by no means a modern phenomena even though its frequency, intensity and spread are now much larger than they were ever before. Volcanic eruptions have been known to cause acid rain even in ancient times. During the last two centuries the increased use of fossil fuels has led to a gradual increase in the emission of sulphur and nitrogen oxides and hence a corresponding increase in the rainfall acidity. This process was hastened by the spurt in industrialisation in the twentieth century and the exponential growth in fossil fuel consumption. There are records of changing diatom (microscopic unicellular algae) population in the lakes of Galloway, Scotland, which reflect an increase in acidity of the lake water during the last two centuries due to acid rain. There was little change in the diatom flora of this lake until the mid-nineteenth century, when diatoms characteristic of neutral water, such as *Brachysira vitrea* began to decline and were gradually replaced by more acidophilous species. Acidification continued through the twentieth century and by the early 1980s the diatom flora of the lake was dominated only by acid tolerant taxa such as *Tabellaria quadriseptata* and *T. binalis*. The lake had a pH of about 4.8 (acidic) in 1980. There is also historical evidence of the disappearance of bogmoss (*Sphagnum* sp.) from the Pennines at the time of Industrial Revolution, as bogmoss is very susceptible to SO_2. Many species of lichens are intolerant to sulphur dioxide and the absence of lichens in industrial areas in Britain indicate a likely impact of acid rain falling in the region over the last two centuries. Nordic countries are among the regions seriously affected by acid rain (Moiseenko 1994).

ORIGIN OF ACID RAIN

Process of formation of acid rain

Precipitation removes gases and particles from the atmosphere by two processes: (i) *rain-out* which is the incorporation of particles into cloud drops which fall to the ground, and (ii) *washout* which occurs when materials below the cloud is swept down by rain or snow as it falls. Pollutants may also undergo direct contact or gravitational settling which is termed as *dry precipitation*. After deposition, acidic products may be neutralized by alkaline soils or carried by seepage, and runoff into lakes, thus contributing towards lowering the pH of the water and affecting the aquatic ecosystem.

Causes of acid rain

There are three main compounds that cause acidification of rain in the atmosphere. They are:

1. *Sulphur compounds and radicals*
   - Sulphur dioxide (SO_2)
   - Sulphur trioxide (SO_3)
   - Hydrogen sulfide (H_2S)
   - Sulphate ions (SO_4^{2-})
   - Sulphuric acid (H_2SO_4)

2. *Nitrogen compounds and radicals*
   - Nitric oxide (NO)
   - Nitrous oxide (N_2O)
   - Nitrogen dioxide (NO_2)
   - Nitrates (NO_3^-)
   - Nitric acid (HNO_3)

3. *Chlorine and hydrochloric acid*
   - Chlorine (Cl_2)
   - Hydrochloric acid (HCl)

Besides these there are other acids which are often responsible for the acidity of rain-water. *Phosphoric acid* mists belong to this category; but the contribution of acids other than the ones mentioned above is not significant. In regions of tropical rain forests, *formic acid* (resulting from forest fires before the rainy season) may cause occasional acid rain.

Substances such as ammonia, calcium carbonate and magnesium carbonate that neutralize acids are also present in the atmosphere, and the 'net' acidity of the rain depends on the chemicals suspended in the air through which the rain falls. In addition to these principal ions bicarbonate (HCO_3^-), carbonate (CO_3^{2-}), potassium (K^+) and sodium (Na^+) ions also play a part in influencing the pH of the rain.

Sources of compounds causing acid rain

Some of these acidifying chemicals are natural and others man-made. In some situations, such as in regions of volcanic eruptions (which emit sulphur and nitrous oxides, SO_x and NO_x) or forest fires (which lead to formic acid as briefly explained earlier), natural sources can play an important role. However, the contribution of anthropogenic sources outweighs that of natural ones in the acidification of rain.

Sulphur oxides: The three main natural sources of sulphur oxides are seas and oceans, volcanic eruptions, biological processes in the soil e.g., the decomposition of organic matter by micro-organisms. The supply of sulphur from these sources is irregular, both in time and space. Man-made sources of emission of sulphur are burning of coal (contributes 60% of SO_2) and petroleum products (contributes 30% of SO_2), and industrial production of sulphuric acid in metallurgical and chemical industries. About 100 million tonnes per year is released by these activites.

Nitrogen oxides: Natural sources of nitrogen oxides are lightning, volcanic eruption, and biological activity, while the anthropogenic sources are power stations, vehicle exhausts, and industries. Together they emit about 90 million tonnes of nitrogen compounds per year.

SO_x and NO_x emissions

SO_x and NO_x emissions are affected by myriad factors. Nearly all of the sulphur present in the fuel in *coal-fired boilers* is converted to SO_2 and is emitted in to the atmosphere along with other fuel gases. *Oil-fired boilers* also emit significant concentrations of sulphur oxides due to the presence of sulphur in fossil oil. Due to this, low-sulphur sources of oil, such as the ones available in Abu Dhabi, are termed 'sweet' and are preferred over oils with higher sulphur content. The presence of *alkaline vapours* such as ammonia (NH_3) in the atmosphere, neutralizes the sulphuric/nitric acids, thereby increasing the pH of the rain or snow. Natural or man-induced *dusts* from wind-blown soil are generally alkaline and may react with and neutralize strong acids in the atmosphere. Coal-fired fly-ash which is alkaline in nature may also help in the neutralization process.

Industries and processes which typically lead to emission of sulphuric, nitric and hydrochloric acids are shown below.

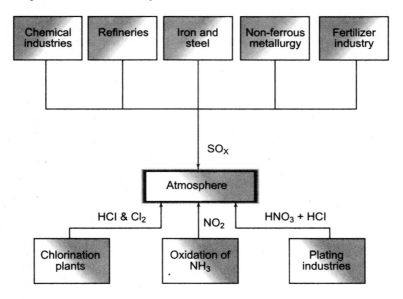

Fig. 2.1 Industries contributing to atmospheric pollution

Refineries: Sulphur is a natural substance present in all crude oils in the form of hydrogen sulphide, mercaptans, thiophenes and polysulphides. Refinery processes lead to the emission of as much as 3500 ppm of sulphur dioxide in fuel gases. Fuel burning also produces nitrogen oxides but refinery processes produce only 2 per cent of the total nitrogen oxides emitted, whereas heaters and boilers account for a much higher proportion.

Iron and steel: Although the main pollutants in this industry are particulates and dust, gas cleaning with oxygen inlet is effected at different stages in the process in order to remove sulphur. This results in SO_x emissions.

Non-ferrous metallurgy: Non-ferrous smelter operations based on sulphide feeds, generate large quantities of sulphur dioxide admixed with air or fuel products.

Chemical industries: Sulphuric acid (H_2SO_4) is a potential raw material for any industry and is manufactured in large quantities. A typical 500 tonnes/day H_2SO_4 plant discharges about 6.5 tonnes/day of SO_2 which is slightly more than the amount of SO_2 released from fuel burning in a 200 MW power station in one hour. Acidic fumes are liberated into the environment as in non-ferrous metallurgical industries. In addition to SO_2 and SO_3, sulphuric acid mist is also formed.

Nitric acid (HNO_3), another important raw material also causes some gas discharges into the environment. It is often produced by the oxidation of ammonia over platinum-rhodium catalysts and the resulting NO_2 is absorbed in water. The gases emerging from smoke stacks usually contain nitrogen (95%), oxygen (3%) and NO_x (500–5000 ppm) often resulting in a brown plume. In UK a limit of 1000 ppm for stack-gas NO_2 was set, whereas EPA has specified 200 ppm.

Another major industrial acid, namely hydrochloric acid, is produced in all chlorination plants (such as those producing vinyl chloride and ethylene dichloride). Chlorine is another gas which escapes into the environment from several industrial areas. Phosphoric acid mists also escape into the atmosphere during phosphoric acid manufacture. These can be completely removed by employing the same mist eliminates used in sulphuric acid manufacture.

Fertilizer industries: In fertilizer plants, a lot of ammonium and diammonium phosphates are manufactured. The usual pollutants that are emitted into the atmosphere by the fertilizer industry are sulphur oxides, sulphuric acid, ammonia and hydrogen chloride.

Plating industries: Metal leaching operations for copper and other metals emit nitric acid fumes into the environment. Hydrochloric acid is often used in metal cleaning stages prior to electroplating and since its vapour pressure is rather high, it is always dispelled into the atmosphere, especially at high temperatures.

In electrolytic applications such as chromium plating, anodising and descaling operations, hydrogen is generated at the cathode and oxygen at the anode. The freshly generated gases escape from the solutions with great speed along with significant quantities of chromic acid and sulphuric acid droplets, thereby contributing to atmospheric pollution.

Miscellaneous: Acidic substances originating from synthetic textile, paper, food, pharmaceuticals, aerospace, and ordnance industries also contribute to the release of acidic gases into the atmosphere.

The spatial distribution and concentration of these emissions may vary depending on the population and the industrialization of the region. It has been estimated that 90 per cent of fossil fuel consumption is in the northern hemisphere.

Evidence of a tenfold increase in atmospheric nitrate concentrations in North America and a fivefold increase in Europe since the turn of the century has come from the analyses made at the agricultural stations in North America and Europe. There is also indirect evidence preserved in the chemical impurities locked up in polar snow and ice. It is possible to determine the exact age of different levels of pollutants within the ice matrices by radioanalytical techniques. Recent analyses of ice-cores from Greenland indicate that atmospheric concentrations of sulphate and nitrate were variable but low before the turn of the century. But, since 1900, they have increased exponentially—nitrates have doubled and sulphates have trebled. The Greenland studies also indicate the widespread distribution of nitrogen oxides caused by aerial dispersion.

Zone of influence and aerial dispersion

In general, concentrations of pollutants in the atmosphere which cause acid rain tend to decrease exponentially with distance from the source to unobjectionably low levels within 100 km. However, if several sources are aligned roughly along the direction of prevailing winds, cumulative concentrations can build up and remain high for substantial distances. These concentrations can be large enough to overwhelm the capacity of the atmosphere to disperse or neutralize them. The long-range transport of air pollutants relates to this cumulative and persistent effect, as well as to the atmosphere's ability at times to transport large volumes of air for long distances with relatively little mixing of (or dilution by) clean air. Radioactive tracer studies of dispersion of polluted air, by various direct and indirect means, have suggested that a *zone of influence* of large sulphur and nitrogen oxide sources can persist for hundreds of kilometres downstream. This long-range cumulative effect creates the potential for relatively high levels of acid deposition over the distances exceeding 1000 km, affecting pristine areas far from industrial-urbanized environments. Thus, the phenomenon of acid rain is generally not identified with localized air pollution problems in and around large sources, or at urban areas, but with larger scale regional effects.

CHEMISTRY OF ACID RAIN

Six basic steps are involved in the formation of acid rain:

1. The atmosphere receives oxides of sulphur and nitrogen from natural and man-made sources.
2. Some of these oxides fall back directly to the ground as *dry deposition*, either close to the place of origin or some distance away.
3. Sunlight stimulates the formation of photo-oxidants (such as ozone) in the atmosphere.
4. These photo-oxidants interact with the oxides of sulphur and nitrogen to produce H_2SO_4 and HNO_3 by oxidation.
5. The oxides of sulphur and nitrogen, photo-oxidants, and other gases (like NH_3) dissolve in the droplets of clouds and rain to produce acidic cations (hydrogen ions, H^+ and ammonium ions, NH_4^+), sulphates (SO_4^{2-}) and nitrates (NO_3^-).
6. Acid rain containing ions of sulphate, nitrate, ammonium and hydrogen falls as *wet deposition*.

The various ways of introduction, transportation, dispersion and deposition of the pollutants on the earth's surface are illustrated in figure 2.2.

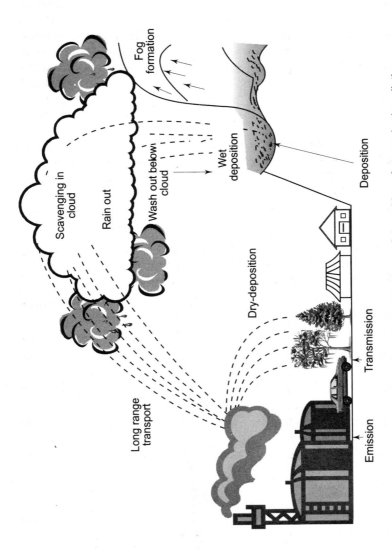

Fig. 2.2 Emission, transport and mechanisms of deposition of atmospheric pollution

Chemical reactions involved in acid rain

Oxidation of sulphur dioxide: In the atmosphere sulphur dioxide gets oxidized to sulphur trioxide which reacts with water or alkalies to give sulphuric acid or sulphates. The acid or sulphates, in particular ammonium sulphate and bisulphate, occur as aerosols. The process is illustrated by the following equation: -

$$SO_2 \xrightarrow{(O)} SO_3 \xrightarrow{(H_2O/NH_4OH)} H_2SO_4/(NH_4)_2 SO_4/NH_4 HSO_4 \quad \ldots (1)$$

The oxidation step proceeds in one or more of three ways, namely, catalytic oxidation, photochemical oxidation and oxidation by radicals.

(1) Catalytic oxidation

$$2SO_2 + 2H_2O + O_2 \longrightarrow 2H_2SO_4, H = 600\,KJ \quad \ldots (2)$$

$$[H \text{ is the heat of the reaction}]$$

This reaction is slow in clean air but is catalyzed by aerosols containing metal those of Manganese ions like (Mn^{2+}) and Iron (Fe^{3+}). Surfaces such as buildings may also act as catalytic centres. The reaction is swifter in places where the relative humidity is more than 32 per cent. The solubility of SO_2 in water is a function of pH and can be described by the equilibria:

$$SO_2(g) + H_2O \rightleftharpoons SO_2(aq) \quad \ldots (3)$$

$$SO_2(aq) + H_2O \rightleftharpoons H_2SO_3 (aq) \quad \ldots (4)$$

$$H_2SO_3 + H_2O \rightleftharpoons H_3O^+ + HSO_3^- \quad \ldots (5)$$

$$HSO_3^- + H_2O \rightleftharpoons H_3O^+ + SO_3^{2-} \quad \ldots (6)$$

$$2HSO_3^- \rightleftharpoons S_2O_5^{2-} + H_2O \quad \ldots (7)$$

$$[g\text{—gaseous, aq—aqueous}]$$

The sulphur species in water is either covalent or ionic, the latter being more soluble due to the high dielectric constant of water.

(2) Photochemical oxidation

SO_2 absorbs solar radiation and converts into its excited states, which then react with oxygen by a variety of routes to give SO_3.

(3) Oxidation by free radicals
Hydrocarbons and nitrogen dioxides, both increase the rate of oxidation of SO_2 to free radicals. These are mostly components of automobile emissions or secondary products.

$$SO_2^{\cdot} + O_3 \longrightarrow SO_3 + O_2 \quad \ldots (8)$$

This reaction is slow in the gas phase and rapid in solution

$$SO_2 + NO_2 \longrightarrow SO_3 + NO \quad \ldots (9)$$

Other feasible reactions are

$$SO_2 + HO_2^{\cdot} \longrightarrow SO_3 + HO^{\cdot} \quad \ldots (10)$$

$$SO_2 + RO_2^{\cdot} \longrightarrow SO_3 + RO^{\cdot}$$

$$\text{(where R indicates } CH_3C = O) \ldots (11)$$

$$HO^{\cdot} + SO_2 \longrightarrow HOSO_2^{\cdot} \quad \ldots (12)$$

$$HOSO_2^{\cdot} + O_2 \longrightarrow HOSO_2O_2 \quad \ldots (13)$$

$$HOSO_2O_2 + NO \longrightarrow HOSO_3 + NO_2 \quad \ldots (14)$$

$$\downarrow\uparrow$$

$$H_2SO_4$$

The dissolution of SO_3 in water gives sulphuric acid which can reduce the pH of the rain to as low as 2. Figure 2.3 is a pictorial representation of the major reactions in the formation of acid rain.

DISTRIBUTION OF ACID RAIN

The geography

Acid rains and the oxides that create them are often transported to distances far away from their points of origin by the wind so that the adverse effects of pollution are experienced at places remote from the place of genesis. This gives rise to problems in handling the resultant pollution as the search for the culprit may well turn out to be a wild goose chase. The problem is further compounded as the environmental damage caused by acid rain is not uniform, but is area-specific, depending upon the geographical and geochemical aspects.

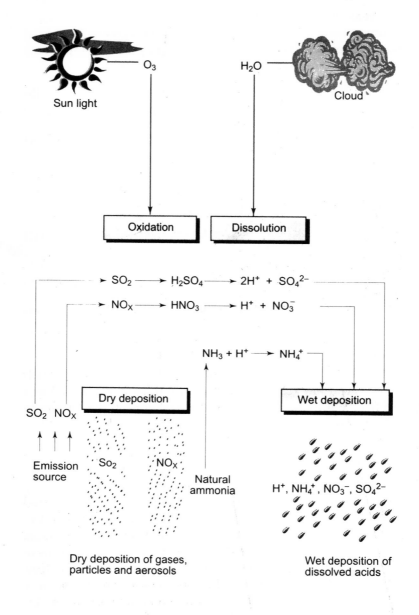

Fig. 2.3 Chemistry of acid precipitation

Areas which are prone to acid-rain attacks have some common characteristics:

- they are concentrated in the industrialized belt of the northern hemisphere
- they are often upland and/or mountainous areas, which are well-watered by rain and snow
- due to the abundance of water, they possess numerous lakes and streams and also have more land covered with vegetation
- being upland, they often have thin soils and glaciated bedrock.

Many parts of Scandinavia, Canada, the North and Northeast United States and North Europe (particularly West Germany and upland Britain) share these features. Across the Atlantic there are a number of acid rain 'hot spots' including Nova Scotia, southern Ontario and Quebec in Canada, the Adirondack Mountains in New York, Great Smoky mountains, parts of Wisconsin, Minnesota, and the Colorado Rockies of the US (La Bastille 1981).

Recent measurements indicate that rains and snows with a pH of 4–3 or lower fall regularly over many of the highly industrialized areas of the northern hemisphere, especially North America and Northern and Western Europe (table 2.1). The intensity of acid rain in European countries is presented in table 2.2. The pH of precipitation in Southern Scandinavia is generally 4.2–4.3. South-east Canada has had rains with a pH range of 4.1–4.6. Many parts of North America have acidic precipitation with a pH of 3.5–4.2. In Britain, rains between 1978 and 1980 had a pH ranging from 4.0 to 4.3. Observations are also now available on the pH of rain in some other parts of the world (Hidey 1995).

The worst hit areas with a pH below 6 are southern Norway, Sweden, Adirondack mountains in New York, southeastern Ontario and some parts of eastern Canada. Southern Norway completely lost its fish population in 20 per cent of the 700 lakes by the late 1970s. The Adirondack mountains have lost the fish populations in 90 per cent of its lakes.

A brief account of what is happening in Poland may sum up the situation. The Polish coal, called 'black gold' due to its role

Table 2.1 Acidification of rain in developed countries

Location	Past pH	Present pH
South-east England	4.5 – 5.0 (1956)	4.1 – 4.4 (1978)
Eastern Scotland	NA	4.2 – 4.4 (1978–80)
Netherlands	4.5 – 5.0 (1956)	NA
Southern Norway	5.0 – 5.5 (1956)	4.7 (1977)
Southern Sweden	5.5 – 6.0 (1956)	4.3 (1975)
Japan (around Tokyo)	NA	4.5 (1983)
Canada (Quebec)	NA	4.5 (1982)
Black Forest (Germany)	NA	4.25 (1972)

Table 2.2 Acid rain intensity in European countries

Country	Level (g) of sulphur per sq metre (annual average)
Netherlands	2.98
Belgium	3.31
Britain	2.61
Portugal	0.73
Spain	1.17
Luxemburg	2.32
France	1.14
Switzerland	1.67
Germany	2.54
Denmark	1.74
Norway	0.67
Sweden	0.68
Finland	0.60
Czech & Slovak Republic	5.20
Poland	4.01
Hungary	3.29
Austria	2.22
Italy	1.71
Yugoslavia	1.98
Romania	1.40
Bulgaria	1.94
Greece	0.81
Turkey	0.25
Albania	0.97

in Polish industry, contains 1–4 per cent sulphur. An estimated 3 million tonnes of sulphur dioxide are reportedly emitted into Poland's polluted atmosphere every year as a consequence of the use of 'black gold'. Together with 'imported' sulphur dioxide from abroad, about 4 million tonnes of SO_2 falls to the ground in Poland every year, amounting to an average value of about 10 tonnes of sulphur compounds per square kilometre area. This figure puts Poland, a relatively small country, in the seventh place of total SO_2 generation after the US, Russia and the former Soviet Union, Canada, Great Britain and the Democratic Republic of Germany. The effects of such overdoses are horrifying. SO_2 fumes with other pollutants are reportedly either causing or contributing to about 600 additional deaths per 100,000 inhabitants every year. People living in Poland are vulnerable to the increased rates of respiratory and circulatory diseases, allergies and leukemia. To make matters worse, scientists speculate that by 1990 about 7.3 million tonnes of sulphur dioxide will be emitted into the atmosphere by Polish factories, power plants and home heating systems that will still burn coal.

Acid rain and developing countries

Although acid rain was first noticed in developed countries and levels and impacts of acid rain in those regions have been extensively studied, instances of acid rain and its harmful impacts are beginning to become common in developing countries as well. Harte (1983) found that the pH of rain falling over Amne Manchim town in China was as low as 1.3, probably the most acidic rain ever recorded. Episodes of acid rain have been seen in Thailand, Philippines, Pakistan and Indonesia (Abito 1995 personal communication, Sarmasek 1995 personal communication).

In India, the first report of acid rain came from Bombay in 1974. Instances of acid rain are being reported from metropolitan cities. A typical study conducted at the heavily industrialised Udyogamandal area in Kerala state revealed that it often receives acid rain; the pH going as low as 4.3 (Khemani et al. 1992). A study conducted at Trombay, Bombay, revealed a high incidence of acidic and sulphate components in air (Khemani et al. 1992).

In India, the annual SO_2 emission has almost doubled in the last decade due to increased fossil fuel consumption, and lowering of soil pH is reported from north-eastern India, coastal Karnataka and Kerala, parts of Orissa, West Bengal and Bihar. Levels of noxious fumes and dust are more than twice the permissible levels in Agra; this is corroding the marble of the well-known monument Taj Mahal, causing it to lose its grace and glitter. This phenomenon has been described as 'marble-cancer'.

Safe areas

There are two types of safe areas where acid rain is not a problem. One comprises areas that simply do not receive acid rain or the gaseous oxides of sulphur and nitrogen due to their location. Almost all of the southern hemisphere is thus protected, as is most of the tropics and parts of the northern hemisphere. The other type is areas that receive acid precipitation but can withstand it. Many areas provide a natural resistance (buffering) to acidification with counter action offered by alkaline soils or limestone beds which neutralize acid inputs. Other forms of buffering are offered in the mid-western United States, where alkaline dust blown from the west neutralizes acid rain before it reaches the ground. Several regions in India also have good buffering capacity but if such regions are persistantly showered with acid rain, the soil may gradually lose its neutralizing ability.

IMPACT OF ACID RAIN

The process of acidification often remains undetected until damage has occurred. In some instances, organisms which are acid sensitive may serve as indicators of the initial stages of acidification. For example, lichens serve as good bio-indicators for air pollution. In the vicinity of pH 6.0, several animals, that are important food items for fish, decline these include the freshwater shrimp (*Gammarus lacustris*), shrimps like *Mysis relicta*, crayfish, snails and some small mussels.

The wide-ranging impacts of acid-rain on the ecosystem includes the effects on its components, as enumerated below.

Freshwater aquatic ecosystems

The effect of acid rain on the chemistry of surface water is determined by the geochemistry, geomorphology and hydrodynamics of the watershed and the waterbody. These factors determine the capacity of water to neutralize acids, and absorb or release metals.

In a group of lakes in southern Norway surveyed in 1933–41 and resurveyed in 1971–5, the pH declined by 0.8 to 1.8 units (Gjessing et al. 1976). Many of these lakes now have pH < 5.5 (Wright and Henrikson 1978). Watt et al. (1979), who resurveyed 19 lakes in Nova Scotia that were originally surveyed in 1955, found that the pH had declined in all of them.

In the Adirondack Mountain regions of New York, 51 per cent of the lakes above 610 m elevation had a pH of less than 5.0 in 1975; when 40 of these lakes were surveyed during 1929–37 only 2 (5 per cent) had a pH of less than 5.0 (Schofield 1976). Lakes that have a pH of ≤ 5.5 and have low concentrations of dissolved organic matter are believed to be highly susceptible to acid precipitation. Before we proceed we must clarify that there *are* a large number of natural lakes which have a pH of less than 5.5 but such naturally acidic lakes have several features different from the lakes which have become acidic due to the receipt of acid rain within a short span of time (3–4 years or less). The difference is summarized in table 2.3.

In natural water bodies metals such as zinc, cadmium, nickel, manganese, chromium, copper and mercury form complexes with naturally occurring ligands such as fulvic acids and phosphates. In the complexed form the bioavailability, and consequently the toxicity, of these metals is greatly reduced (Abbasi 1995; Soni 1991; Abbasi & Soni, 1983, 1986, 1993). pH also effects the mobility of metals present in sediments, towards the overlaying water through direct dissolution; the solubility of most metal compounds in water decreases as the pH increase (Abbasi et al. 1988, 1992, 1994). When a lake gets acidified, the metal-fulvic acid complexes are destabilized releasing metal ions. A reduction in pH also increases mobility of the metal compounds from the sediments to the overlaying water. The combined effect of these two phenomena increases the concentration of metals in water. The concentration often increases to levels that causes chronic toxicity and even acute toxicity in aquatic organisms (Abbasi 1995, 1991; Soni 1991; Abbasi & Abbasi 1995; Abbasi et al. 1988, 1994, 1995).

Acidified lakes in Norway, Sweden and Canada (Ontario) have higher concentrations of zinc, lead, copper, cadmium and nickel than similar lakes in non-acidified areas, and these metals

Table 2.3 Characteristics of naturally acidic lakes and anthropogenically acidified natural lakes

Naturally acidic lakes	Anthropogenically acidified lakes
Brown to yellow colour caused by humic substances (peat, tannins, etc.)	Very clear water caused by reduced primary productivity (ie, lack of phytoplankton and consequently zooplankton), precipitation of organic matter by aluminium and dissolution of iron and manganese colloids. Inorganic acids (particularly sulphuric and nitric acids) predominate in these lakes.
Concentrations of dissolved organic carbon are high while transparency is low.	Dissolved organic carbon concentrations are low, (seldom exceeding a few milligrams per litre) whereas the transparency is high.
Low pH (below 5) but well buffered: these lakes generally have a greater acid neutralizing capacity (by organic and inorganic buffers, and the action of micro-organisms). This suggests that brown-water lakes would be less sensitive to the acidifying effects of precipitation.	Poorly buffered: once the bicarbonate alkalinity is exhausted, the pH decreases rapidly, giving little time for the species to adapt.
Abound with aquatic life.	Some of the more sensitive taxa, such as blue-green algae, some bacteria, snails, mussels, crustaceans, mayflies and fish either decrease or are eliminated.
The presence of humic, fulvic, and other organic acids keeps the availability of toxic metals in check.	No such cushion available.

have been detected in precipitation in these areas (Wright & Gjessing 1976).

Acidified lakes are most likely to be found in regions where the bedrock and soil are low in naturally occurring acid-neutralizing materials. In regions where limestone is abundant, acid precipitation gets neutralized easily.

Acidification has an adverse effect on all trophic levels of aquatic biota in surface water, from decomposers to fish and amphibians (table 2.4). Studies in acidified surface waters, and *in vitro* laboratory experiments have shown that microbial activity is reduced at low pH, and the dominance of species shifts from bacteria to fungi (Leivestad et al. 1976, Andersson et al. 1978).

As lakes become acidified, the diversity of phytoplankton

Table 2.4 Effects of decreasing pH on aquatic organisms

pH	Effect
8.0–6.0	A decrease of less than one-half of a pH unit in the range of 8.0 to 6.0 is likely to alter the biotic composition of lakes and streams to some degree in the long run. However, the significance of these slight changes is not great.
	A decrease of one-half to one pH unit (a threefold to tenfold increase in acidity) may detectably alter community composition. Some species may be eliminated.
6.0–5.5	Decreasing pH from 6.0 to 5.5 will reduce the number of species in lakes and streams. In the remaining species, significant alterations in the ability to withstand stress may occur. Reproduction of some salamander species is impaired.
5.5–5.0	Below pH 5.5, numbers and diversity of species will be reduced. Reproduction is impaired and many species will be eliminated. Crustaceans, zooplankton, phytoplankton, molluscs, amphipods, most mayfly species, and many stone-fly species will begin to die out. In contrast, several invertebrate species tolerant to low pH will become abundant. Overall, invertebrate biomass will be decrease greatly. Certain higher aquatic plants will be eliminated.
5.0–4.5	Below pH 5.0, decompostion of organic detritus will be impaired severely. Most fish species will be eliminated.
4.5 and below	In addition to exacerbation of the above changes, many forms of algae will not survive at a pH of less than 4.5.

species declines due to stress on acid-sensitive organisms. As a result the entire composition of species changes. At low pH the *Chrysophyta, Cyanophyta* and *Chlorophyta* decrease and *Pyrrhophyta* become dominant (Almer et al. 1974, Yan 1979). The number of species of zooplankton present in a water body was found to decrease as pH decreased, in studies conducted in Sweden, Canada and Norway (Decosta 1980, Hendrey and Wright 1976, Sprules 1975).

Molluscs, crustaceans, ephemeroptera and plecoptera appear to be highly sensitive to acidification. Invertebrates in acidic lakes may accumulate toxic trace elements that could then be passed on to vertebrates in the food web. The most dramatic effects of acidification on aquatic organisms have been on fish. (Abbasi 1995). The observed effects include mortality, reproductive failure and increased heavy metal uptake.

The acidification of aquatic habitats also affects some amphibians. Pough (1976) found that one species of salamander, *Ambystoma macalatum*, was sensitive to acid. Mortality of embryos was high at pH < 6.0.

Soils and Vegetation

Soils of different regions, have differing ability to process and neutralize acidic inputs. Soils have been broadly classified into four categories on the basis of their acid-neutralizing abilities.

Type I—Low or no buffering capacity, overlying waters very sensitive to acidification (granite/syenite, granitic gneisses, quartz sandstones, or equivalents).

Type II—Medium to low buffering capacity, acidification restricted to first and second order streams and small lakes, sandstones, shales, conglomerates, high grade metamorphic to intermediate volcanic rocks, intermediate igneous rocks, calci-silicate gneisses.

Type III—High to medium buffering capacity, no acidification except in cases of overland run-off in areas of frozen ground (slightly calcareous), low grade intermediate to mafic volcanic, ultramafic, glassy volcanic rocks.

Type IV—'Infinite' buffering capacity, no acid precipitation effect of any kind (highly fossiliferous sediments or metamorphic equivalents, limestones, dolostones).

The acid neutralizing ability of the soil of a given region greatly influences the extent of harm acid rain may cause in that region.

The principal pathways and mechanisms of plant growth response to atmospheric deposition and gaseous pollutants are illustrated in figure 2.4.

The exchange between hydrogen ions and the nutrient cations like potassium and magnesium in the soil cause leaching of the nutrients, making the soil infertile. This is accompanied by a decrease in the respiration of soil organisms. An increase in ammonia in the soil due to a decrease in other nutrients decreases the rate of decomposition. The nitrate level of the soil is also found to decrease. The impact of acid rain on soil is more obvious in United Kingdom and other European countries than in India; because Indian soils are mostly alkaline, with good buffering ability.

Acid rain affects trees and undergrowth in forests in several ways, causing reduced growth or abnormal growth:

The typical growth-decreasing symptoms are:

- discoloration and loss of foliar biomass (yellowing and browning of needles and leaves)
- loss of feeder-root biomass, especially in conifers
- decreased annual increment (width of growth rings)
- premature senescence (aging) of older needles in conifers
- increased susceptibility to secondary root and foliar pathogens
- death of herbaceous vegetation beneath affected trees (all within the drip line from the canopy)
- prodigious production of lichens on affected trees.
- death of affected trees.

Abnormal growth symptoms are:

- active shedding of needles and leaves while still green, with no indication of disease
- shedding of whole green shoots, especially in spruce
- formation of stork's nest crown in young white fir
- altered branching habit and greater than normal production of adventitious (out-of-place) shoots
- excessive seed and cone production year after year

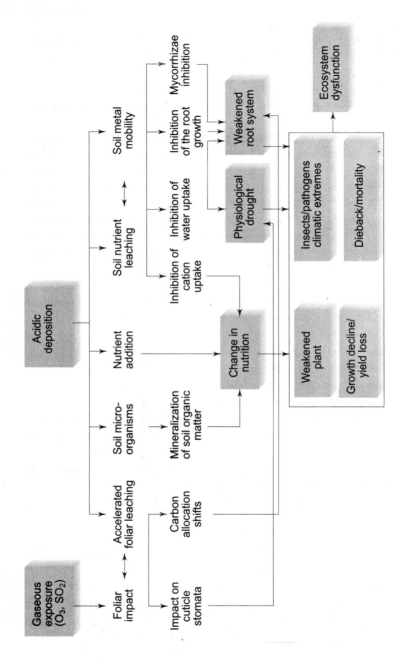

Fig. 2.4 Impact of acid rain on ecosystems

- diseased trees show growth at the top and not near the bottom, with very brittle branches.

In addition water-stress symptoms may also be caused:

- altered water balance
- increased incidence of wet-wood disease.

Acid rain may cause a shift in the community structure of micro-organisms (as explained in the following section); useful micro-organisms in soil may be replaced by disease causing fungi and the population of beneficial earthworms may be reduced or eliminated.

Micro-organisms

pH determines whether any microbial species can proliferate in a particular environment and the rate at which it can reproduce. The pH optima of most bacteria and protozoa is near neutrality; most fungi prefer an acidic environment, most blue-green bacteria prefer an alkaline environment. So after a long run of acid rain, microbial species in the soil and water shift from bacteria-bound to fungi-bound and cause an imbalance in the microflora. This causes a delay in the decomposition of soil organic material, and an increase in fungal diseases in aquatic life and forests. Also a latent burden on water supply treatment systems.

The effect of pH on the transport of material across the membrane of micro-organisms is also a very important factor and perhaps the determining factor in influencing their growth. The ionization of a required nutrient may make it unavailable to the cell, whereas a toxic compound may become inhibitory only at a pH at which it is not ionized.

Wildlife

The effects of acid rain on wild life are not very obvious and are therefore, difficult to document. Nevertheless, several direct and indirect effects of acid rain on the productivity and survival of wildlife populations have been reported.

Acid rain can directly affect the eggs and tadpoles of frogs and salamanders that breed in small forest ponds (Abbasi et al. 1989, 1995). The effects of acid rain are mostly unknown for wildlife that do not depend on the aquatic medium as much as on the total environment for their eggs or young.

It has been postulated that acid rain can indirectly affect wildlife by allowing metals bound in soils and sediments to be released into the aquatic environment, where toxic substances may be ingested by animals, like birds, that feed in such an environment. For example, an accumulation of toxic levels of aluminium in the food chain may explain the reproductive failure observed in flycatchers that feed on insects emerging from acidified lakes in Scandinavia.

Other indirect effects of acid rain on wildlife are loss or alteration of food and habitat resources. The elimination of fish from a lake, for example, may mean that water birds like loons and mergansers which survive on fish might desert the lake for want of food. However, an increase in the emerging insect densities after elimination of fish might benefit other water-bird species during their nesting or brood rearing state.

Degradation of plant communities by acid rain can include damage to leaves, loss of productivity and sometimes loss of sensitive plants. Because abundance and diversity of plants is directly related to that of wildlife, any effect on the former will also affect the latter.

Humans and Buildings

Historical monuments such as the Taj Mahal in India have been affected by what has been termed 'stone-cancer' or 'marble-cancer'. Stone-cancer is yet another effect of acid rain—caused by its corrosive action on the building material—marble. Numerous other materials are harmed by acid rain or the gases responsible for acid rain (table 2.5).

Acid rain affects human health is a number of ways. The obvious ones are bad smells, reduced visibility; irritation of the skin, eyes and the respiratory tract. Some direct effects include chronic bronchitis, pulmonary emphysema and cancer. Some indirect effects include food poisoning *vis a vis* drinking water and food. An increase in the levels of toxic heavy-metals like manganese, copper, cadmium and aluminium also contribute to the detrimental effects on human health. According to a report on the susceptibility of UK groundwaters to acid-deposition-

Table 2.5 Acid rain damage to materials

Material	Type of Impact	Principal Air Pollutants	Other Environmental Factors	Mitigation Measures
Metals	Corrosion, tarnishing	Sulphur Oxides and other acid gases	Moisture, air, salt, particulate matter	Surface plating or coating replacement with corrosion resistant material, removal to control environment
Building stone	Surface erosion soiling, black crust formation	Sulphur oxides and other acid gases	Mechanical erosion, particulate, moisture, temperature fluctuations, salt, vibration, CO_2, micro-organisms	Cleaning, impregnation with resins, removal to controlled environment
Ceramics and glass	Surface erosion, surface crust formation	Acid gases, especially fluoride-containing	Moisture	Protective coating, replacement with more resistant material, removal to controlled atmosphere
Paints and organic coatings	Surface erosion, discolouration, soiling	Sulphur dioxides, hydrogen sulphide	Moisture, sunlight, ozone particulate matter, mechanical erosion, micro-organisms	Repairing, replacement with more resistant material
Paper	Embrittlement, discolouration	Sulphur oxides	Moisture, physical wear, acidic materials introduced in manufacture	Synthetic coatings, storing in controlled atmosphere, deacidification, encapsulation, impregnation with organic polymers

Contd.

Table 2.5 (Contd.)

Material	Type of Impact	Principal Air Pollutants	Other Environmental Factors	Mitigation Measures
Photographic materials	Micro-blemishes	Sulphur oxides	Particulate matter, moisture	Removal to controlled atmosphere
Textiles	Reduced tensile strength, soiling	Sulphur and nitrogen oxides	Particulate matter, moisture, light, physical wear, washing	Replacement, use of substitute materials, impregnation with polymers.
Textile dyes	Fading, colour change	Nitrogen oxides, ozone	Light, temperature	Replacements, use of substitute materials, removal to controlled environment
Leather	Weakening, powdered surface	Sulphur oxides	Physical wear, residual acids introduced during manufacture	Removal to controlled atmosphere, consolidated with polymers, or replacement
Rubber	Cracking	Ozone	Sunlight, physical wear	Add antioxidants to formulation, replace with more resistant materials

induced acidification, the most serious consequence of a significant fall in groundwater pH, was likely to be the increased solubility of metals in potable water and an accelerated corrosion of the distribution network. The greatest threat is probably to small, untreated local supplies from springs, shallow wells and boreholes. Acidic groundwater in shallow wells and at lake inflows have been reported in Sweden.

There is evidence to suggest that acid flushes may be associated with high concentrations of sulphate and nitrate in stream waters. This is particularly the case during snowmelt, when the processes which normally help to regulate the drainage into streams and lakes may be substantially curtailed. For nitrate the problem may be further exacerbated by the soil freeze/thaw effect, which may mobilize even more nitrate (Likens et al. 1974). Ultimately, high nitrate levels could cause a greater water quality problem than high hydrogen ion concentrations. There is strong historical evidence for a dramatic increase in the nitrate component of acid rain in Europe and North America (Likens et al. 1974). When high concentrations of sulphate pass through soils into drainage water, there is often an associated increase in the concentration of the base element and other metal cationic species. Aluminium mobilized in this way may play a major role in the adverse effects of acidic freshwaters on fish and other aquatic life. Sulphate anion adsorption leads to base cation retention, as discussed later, but sulphate saturation of adsorption sites facilitates cation leaching (Cresser et al. 1987).

As acid rain enhances the scavenging of toxic materials from the atmosphere, mobilizes natural or synthetic contaminants in catchment areas, and promotes corrosion of drinking water distribution systems, as indicated above, it would eventually lead to bio-accumulation of toxic chemicals in the human body (Jeng-Peng and Bhargava 1996). In Sweden, groundwater, which is the main drinking water reserve, is already becoming acidic in some areas. The average adult ingests two litres of water each day, including water used to prepare food and beverages. Children and infants consume a disproportionately greater amount with respect to body weight. Thus, babies may be unsafe even when the water they consume contains toxic materials within the permissible limit. In some remote areas where drinking water is obtained locally, trace contaminants mobilized by acid deposition could be an important route to human exposure. This problem is complicated by the fact that these people may not even be aware of the dangers they face.

A low pH of the rainwater and subsequent increased acidity in the environment can trigger off or aggravate the effects of certain harmful pollutants.

Mercury: Methyl mercury and related short chain alkyl mercurial compounds are the environmental contaminants most dangerous to humans, primarily because they accumulate in edible fish tissue. Although acid deposition may not increase the production of methyl mercury it may increase the partitioning of methyl mercury into the water column. In some recent efforts to neutralize acidic lakes, the use of lime appears to have helped in reducing the mercury levels in fish.

Aluminium: Acidified waters are known to leach substantial amounts of aluminium from watersheds. Concentrations of aluminium in acidified well-waters in the USA have been found to be as high as 1.7 mg/L, that far exceeds its background level. Alum-treated water shows aluminium levels three times higher than that in untreated waters. Even at relatively low levels, aluminium has been implicated in dialysis dementia, a disorder of the central nervous system, which may be toxic to individuals with impaired kidney function.

Cadmium: Cadmium can enter the drinking water supply through corrosion of galvanized pipe or from the copper-zinc solder used in the distribution systems. A decrease in water pH from 6.5 to 4.5 can result in a fivefold increase in cadmium and could cause renal tubular damage in 0.9 per cent of the population.

Lead: As is the case with several other metals, the dissolution of lead in water increases with increasing acidity of water. The levels of lead thus elevated in drinking water can be hazardous in many ways. A survey has revealed that drinking water in 16.1 per cent of 2654 rural US households studied had lead levels in excess of the 50 mμg/L maximum acceptable limit. Foetuses and infants are highly susceptible to drinking water lead contamination. Because lead can cross the placenta, foetuses can absorb up to 50 per cent of ingested lead, whereas adults absorb only 8 per cent. High blood lead levels in children (>30 mμg/mL) are believed to induce biochemical and neurophysiological dysfunction. However, lower than normal blood levels of lead can cause mental deficiencies and behavioural problems.

Asbestos: Asbestos in natural rock can be released by acidic waters, to a level of 500 million fibres/L. As many as 40 million people in the US may be exposed to asbestos in their drinking water. The human health effects of increased ingestion of asbestos fibres are not fully understood yet.

General impact of corrosion of distribution systems: When water of low (acidic) pH passes through the water supply pipelines, it corrodes the supply lines and solubilizes metals. As much as 85–92 per cent of the drinking water supplies in the acid-sensitive North-Eastern US were found to be affected by such corrosion. This problem is also serious in Northern Europe.

Socio-economic impacts: The adverse impact of acid rain on farming and fishing leads to the deterioration of life quality indices like GNP and per capita income, especially in the predominantly agricultural and developing countries like India.

ACID RAIN—INTERNATIONAL PERSPECTIVES

The most disputed aspect of acid rain is the 'cleaning-up' process. The clean up programmes will be expensive and might yield only qualified improvements. Inevitably the high costs involved must be met by the countries that export the pollution, whereas the benefits will be enjoyed mostly by the importer countries. Thus arises the inevitable difference of opinion between countries over the best way of tackling the problem of acid rain. Those countries that are hardest hit by the problem (such as Canada & the Nordic States) are convinced that something must be done without further delay, to reduce emissions and thus save their forests and lakes. Twenty-one governments resolved to cut their own emissions of sulphur dioxide by atleast 30 per cent by 1993 and hope to persuade the main producers of the culprit oxides to join them in a truly international solution to the problem. The main contributors to acid rain are the United States and Great Britain. In December 1984, the UK Parliamentary Select Committee on the Environment while reporting on acid rain concluded that Britain has the highest level of air pollution in western Europe and that it should take immediate action to reduce emission of sulphur and nitrogen oxides.

The international scientific community regards acid rain as one of the most serious and significant environmental problems of our time, requiring urgent attention and immediate political initiative. Britain and the United States argue that the problem is as yet too little understood and that links with possible source areas and likely sources such as coal-fired power stations are too ill-defined still to justify wholesale political intervention in the form of the introduction of emission controls which would cause electricity price hikes for the consumer. The acid rain debate is thus a terrain for battle in politics and a struggle in science.

Acid rain was not perceived as a serious threat in India till recently and did not feature as frequently in environment—related debates as other issues of common concern. However, reports on the corrosion of the Taj Mahal helped in focussing public attention on the threats India faced from acid rain. This realization was forcefully brought home when the Supreme Court recently (1994) ordered the closure of the oil refineries and other industries implicated in corroding the Taj Mahal with their acidic emissions. It is evident that the gravity of the situation has sunk in.

CONTROL MEASURES

Most of the effects of acid rain on fish and wildlife resources cannot be directly countered without reducing or eliminating the sources of pollution. However, some measures may be taken locally to counter acidification of streams, soil, and the rest of the environment. Some of the measures are:

Buffering: The practice of adding a neutralizing agent to the acidified water to increase the pH is one of the important control measures. Usually lime in the form of calcium oxide and calcium carbonate is used.

Breeding hatchery fish in acidified water: Fish are bred in an environment with gradually increasing acidity so that they become acclimatized and are able to survive in an acidified habitat. They are then reintroduced into acidified waterbodies that have a lean fish population. This practice replenishes fish population that would otherwise decline, thus helping to maintain an ecological balance within acidified lakes.

Prevention of the emission of excess SO_x and NO_x from industries: Laws should be formulated by the government to regulate the emission. Some of the measures are: i) decreasing emission of SO_2 from power stations by burning less fossil fuel, using alternate energy sources like tidal, wind, hydropower etc; ii) using low sulphur fuel; iii) desulphurization; iv) decreasing emission of NO_x from power stations; and v) modification of engines. Emissions of SO_x can be controlled in various ways:

- converting it to sulphuric acid
- converting it to elemental sulphur
- neutralizing it and using it in the manufacture of other products.

In all these processes, the gas mixture containing sulphur dioxide is first cleaned with water in order to eliminate impurities such as arsenic, selenium salts, mercury, chlorides, and sulphides. The gas stream is then passed through a succession of catalyst beds (usually vanadium oxide, V_2O_3) and then in to an absorption vessel (H_2SO_4, single contact). The efficiency of the conversion of SO_2 to SO_3 in this process is 97–8 per cent, provided the gas strength is greater than 3.5 per cent. The efficiency of conversion to SO_3 can be increased to 99.9 per cent if the residual gas on first sequence of absorption through vessel is reheated and brought once again to the catalyst beds. In order to run this process successfully, the original air mixture leaving the smelter should have a SO_2 concentration higher than 6 per cent and the final stack gas concentration below 500 ppm.

Elemental sulphur can be recovered from the smelter gases by any of the processes described below. The gas concentration should be brought to a low level (up to 1–3 per cent SO_2) and the sulphur produced should be dried, de-dusted and cooled to 50°C.

1. Absorption of gas in buffered citric acid/sodium citrate at pH 3.8 (strictly) and removal of sulphur from the solution by filtration/thickening.

2. Liquid phase Claus-conversion to sulphur.

$$SO_2 + 2H_2S \longrightarrow 3S + 2H_2O$$

The Hydrogen Sulphide (H_2S) required for this is generated by the reaction shown below:

$$CH_4 + 4S + 2H_2O \longrightarrow CO_2 + 4H_2S$$

The typical costs of SO_x removal are given in table 2.6.

Acid rain and its effects will simply not vanish without effort. Increasing public awareness of the problem is the first step towards finding some solution to the acid rain problem. The cost of control versus the cost of damage must be considered while evaluating the merit of any management alternative. It is a problem not without a hope. The right strategy can help to improve the situation.

HOW BIG A THREAT IS ACID RAIN?

As is with other environmental issues, the phenomenon of acid rain has been enveloped in controversy. There have been two radical and conflicting sets of opinions—one believing that acid rain would cause catastropic damage to the environment and the other playing down the acid rain hazard to something more or less inconsequential.

When one sifts through the literature available on acid rain, one finds numerous original papers and reviews describing adverse impacts of acid rain in which the authors have used one or other body of evidence to drive home their own belief on the extent of the threat of acid rain. Lying between the two extremes are opinions of varying extents of moderation.

On balancing the available evidence, the following pointers emerge:

• Acid rain is a slow acting scourge—its impacts are not dramatically evident over a short time span unless, in exceptional cases, the acidification of the rain has taken place suddenly and sharply.

- Some regions feel the impact of acid rain more easily than other regions and this essentially depends on the availability, or the lack of, acid-neutralizing dust, soil or water in the region.
- It is simplistic to believe that if in a particular region acidification of rain is not presently causing proportional acidification of receiving water and soil, this will continue

Table 2.6 Representative costs of strategies to reduce SO_x emissions

Control Strategies	Costs (in US dollars per tonne of SO_2)
Coal cleaning	
North Appalachia and east Midwest coal	50–600
South Appalachia coal	700–1000
Utility strategies[a]	
Fuel switching	
Shift from high to low-sulphur coal	250–350
Shift from high to medium-sulphur coal	350–400
Shift from medium to low-sulphur coal	400–500
Shift from high to low-sulphur residual oil	300–400
Fuel gas desulphurization	
Shift from unscrubbed to scrubbed high sulphur coal	400–600
Shift from unscrubbed to scrubbed medium sulphur coal	600–1500
Shift from unscrubbed to scrubbed low-sulphur coal	1800–3000
Limestone injection multistaged burns	
High-sulphur coal	200–350
Medium-sulphur coal	250–700
Low-sulphur coal	500–1200
Industrial strategies[b]	
Fuel switching	
Shift from high to low-sulphur coal	250–350
Shift from high to medium-sulphur coal	350–400
Shift from medium to low-sulphur coal	400–500
Shift from high to low-sulphur residual oil	300–400
Fuel gas desulphurization	
Shift from unscrubbed to scrubbed high-sulphur coal	400–600
Shift from unscrubbed to scrubbed medium-sulphur coal	600–1500
Shift from unscrubbed to scrubbed low-sulphur coal	1800–3000

a For a 500 MW power plant

b For a 170 Million Btu/hr industrial boiler (Btu: British thermal unit)

to be so indefinitely. If not checked acid rain would gradually wear down the acid-assimilative capacity of the receiving environments.

REFERENCES

Abbasi, S.A. (1986) 'Binary and Ternary Complexes of Interest to Environmental Systems'. *Polish Journal of Chemistry* 64: 339–47.

—— (1991) *Environmental Impact of Water Resources Projects*. New Delhi: Discovery Publishing House.

—— (1995) 'Studies in Environmental Analysis'. DSc thesis, University of Mysore.

——, V. Baji and R. Soni (1989) 'Simulation of the Impact of Acid Rain and Alkaline Wastewater Inflow on Pond Biota'. *International Journal of Enviromental Studies* 35: 97–103.

——, T. Kunhamed, P.C. Nipaney and R. Soni (1995) 'Effect of pH on Chromium Toxicity in Acidified Water Bodies'. *Pollution Research* 14: 317–23.

—— and N. Abbasi (1995) *Clean Water: Great Needs, Greater Challenges*. Karad: Enviromedia.

——, N. Abbasi and R. Soni (1998) *Heavy Metals in Environment*. New Delhi: Mittal Publications.

——, P.C. Nipaney and D.S. Arya (1994) 'Heavy Metals in the Sediments of a River Impacted by Pulp-and-Paper Effluents'. *Journal of Institution of Public Health Engineers*, pp. 18–23.

——, P.C. Nipaney and R. Soni (1988) 'Studies on Environmental Management of Mercury, Chromium, and Zinc with Respect to Impact on Some Arthropods and Protozoans'. *International Journal of Environmental Studies* 32: 181–7.

——, P.C. Nipaney, R. Soni and D.S. Arya (1992) 'Assessment of Water Quality for Cobalt, Nickel, and Copper'. *Journal of Institution of Public Health Engineers*, pp. 8–16.

—— and R. Soni (1983) 'Teratogenic Effects of Chromium in Environment as Evidenced by the Impact on Larvae of Amphibian Rana Tigrina'. *International Journal of Environmental Studies* 23: 131–7.

—— and R. Soni (1986) 'An Examination of Environmentally Safe Levels of Zinc, Cadmium, and Lead with Reference to Impact on Channelfish Nuria Denricus'. *Environmental Pollution* A40: 37–51.

—— and R. Soni (1993) 'Computer-aided Studies on Environmental Impact Assessment and Management of Seven Heavy Metals'. *Journal of Institution of Public Health Engineers*, p. 1–4.

Almer, A., B.W. Dickson, E. Ekstrom and E. Hornstrom (1987) 'Sulphur Pollution and the Aquatic Ecosystem'. In *Sulphur in the Environment, Part II, Ecological Impacts*, edited by J. Nriagu. New York: John Wiley & Sons, pp. 273–311.

Andersson G., S. Fleischer and W. Graneli (1978) 'Influence of Acidification on Decomposition Processes in Lake sediment verh.' Int. Verein, Limnol 20: 802–7.

De Costa, J. (1980) 'The Zooplankton Communities of Acidic Lakes'. Abstracts of voluntary contributions, International conference on the ecological impact of acid precipitation, San Defjored, Norway, March 11–14.

Gilbert, F. (1975) 'History of Acid Rain'. In *Acid Rain: Theoretic and Reality*, edited by C. Park. New York: John Wiley & Sons.

Gjessing, E., A. Henriksem, M. Johannessen and R.Wright (1976) 'Effects of Acid Precipitation on Fresh Water Chemistry'. In *Impact of Acid Precipitation on Forest and Freshwater Ecosystems in Norway*, edited by F. Brackke, Acid Precipitation—Effects on Forest and Fish Project, Aas, Norway, Research Report–6, pp. 64–85.

Gorham, E. (1955) 'On the Acidity and Salinity of Rain'. *Geochim et Cosmochim Acta* 7.

Harte, J. (1983) 'An Investigation of Acid Precipitation in Quighai province, China'. *Atmospheric Environment* 17.

Hendrey, G. and R. Wright (1976) 'Acid Precipitation in Norway, Effects on Aquatic Fauna'. *Journal of Great Lakes Research* 2 (supplement 1): 192–207.

Hidey, G. (1995) 'Acid Rain'. In *Encyclopedia of Environmental Biology*. New York: Academic Press, pp. 1–17.

Jeng-Peng, L. and D.S. Bhargava (1996) 'The Menace of Acid Rain'. *Ecology* 10(9): 19–34.

Khemani, L.T., G.A. Momin, P.S. Prakasa Rao, P.O. Safai, G. Singh and R.K. Kapoor (1992) 'Spread of Acid Rain Over India'. *Atmospheric Environment* 23.

La Bastille, A. (1981) 'Acid Rain: How Great a Menance'. *National Geographic* 160.

Leivestad, H., G. Hendrey I. Muniz, and E. Snetki (1976) 'Effects of Acid Precipitation on Freshwater Organisms'. In *Impact of Acid Precipitation on Forest and Fresh Water Ecosystems in Norway,* edited by F. Brackke, Acid Precipitation—Forest and Fish Project, Aas, Norway, Research Report–6, pp. 86–111.

Likens, G. and F. Bormann (1974) 'Acid Rain: A Serious Regional Environmental Problem'. *Science* 184: 1176–79.

Moiseenko, T. (1994) 'Acidification and Critical Loads in Surface Waters: Kola, Northern Russia'. *Ambio* 23(7): 418–34.

Odan, S. (1976) 'The Acidity Problem—An Outline of Concepts'. *Water, Air, Soil Pollution* 6: 317–65.

Park, C. (1987) *Acid Rain: The Rhetoric and Reality*. New York: Methuen.

Pough, F. (1976) 'Acid Precipitation and Embryonic Mortality of Spotted Salamanders'. *Science* 192: 68–70.

Sanhueza E., Z. Ferrer, J. Romero and M. Santana (1991) 'HCHO and HCOOH in Tropical Rains'. *Ambio* 20(3): 115–17.

Schofield, C. (1976) 'Acid Precipitation: Effects on Fish'. *Ambio* 5.

Soni, R. (1991) 'Impact of Pollutants on Aquatic Organisms'. Ph.D. thesis, University of Calicut.

Sprules, W. (1975) 'Midsummer Crustacean Zooplanton Communities in Acid-stressed Lakes'. *Journal Fisheries Research Board Canada* 32.

Watt, W., D. Scott and S. Ray (1979) 'Acidification and Other Chemical Changes in Halifax County lakes after 21 years'. *Limnology and Oceanography* 23.

Wright, R. and E. Gjessing (1976) 'Acid Precipitation: Changes in the Chemical Composition of Lakes'. *Ambio* 5.

Wright, R. and A. Henriksen (1978) 'Chemistry of Small Norwegian Lakes, With Special Reference to Acid Precipitation'. *Limnology and Oceanography* 23.

Yan, N. (1979) 'Phytoplankton Community of an Acidified, Heavy-metal Contaminated Lake near Sudbury, Ontario: 1973–1977'. *Water, Air, Soil Pollution* 11.

3

The Ozone Hole

To a layman the term ozone hole may conjure up visions of a hole in the atmosphere overhead, deepening and enlarging with each passing day. But has a hole actually developed in the ozone layer? If so, how is it going to hurt us? Is it just a scare? Or is the wolf really at the door?

DIFFERENT ATMOSPHERIC LAYERS

Before we discuss what ozone is and how its 'layer' in the atmosphere gets punctured, let us discuss the constituents of the atmosphere. The atmosphere consists of the following regions or 'layers':

Troposphere: This is characterized by a decrease in temperature with increasing altitude. Around the equatorial regions, the troposphere extends up to 17 km and is only 6 to 8 km deep at the poles. The upper limit of the troposphere is called 'the tropopause'.

Stratosphere: This region extends up to 50 km in altitude. It is characterized by an increase in temperature with increasing altitude. This is due to the ability of the ozone gas, which is present in this region, to absorb sunlight. The increase in temperature becomes more rapid with increasing altitude within the upper parts of the stratosphere. The temperature can reach a maximum of about 270°K. The upper limit of the stratosphere is called 'the stratopause'.

Ozone is found primarily in the stratosphere within the 10–50 km range in altitude. This stretch is referred to as 'the ozonosphere'.

Mesosphere: This region extends from 50 km up to 85 km in altitude. It is characterized by a rapid decrease in temperature with increase in altitude. A minimum of 160°K can be reached at high altitudes. The upper limit of the mesosphere is called 'the mesopause'.

Thermosphere: This region is characterized by a continuous increase in temperature. The temperature can attain a value of 500°K in the course of a night or during minimal solar activity and to above 1,750°K in the course of the day during maximum solar activity. 'Thermopause' refers to the upper limit of the thermosphere; beyond it is the isothermal region.

Figure 3.1 shows the different atmospheric layers and figure 3.2 depicts the altitudes of their distribution.

Chemically the atmosphere may be understood as consisting of two main regions:

Homosphere: This extends up to 100 km and consists of a mixture of nitrogen and oxygen.

Heterosphere: This extends beyond 100 km, where molecular oxygen is strongly dissociated and atomic oxygen is an important

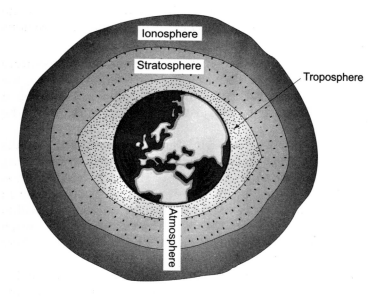

Fig. 3.1 Layers of the atmosphere (horizontal cross-section)

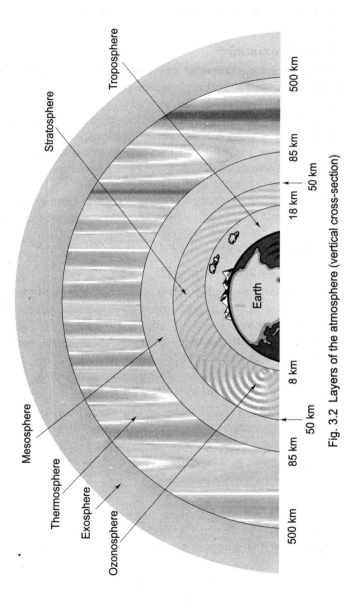

Fig. 3.2 Layers of the atmosphere (vertical cross-section)

constituent of the atmosphere. Helium and hydrogen are also found in this region.

What is ozone?

Ozone is an allotrope of oxygen consisting of three atoms of oxygen bound together in a non-linear fashion. The chemical symbol of ozone is O_3.

The configuration of the ozone molecule and its chemical properties are such that ozone efficiently absorbs ultraviolet light, thus acting like a sun-screen. In doing so, ozone protects oxygen at lower altitudes from being broken up by the action of ultraviolet light and also keeps most of the ultraviolet radiation from reaching the earth's surface. Hence, ozone plays a significant role in protecting the environment even though it constitutes less than one part per million of the gases in the atmosphere (Thompson 1992).

How was ozone formed?

Evolutionary geologists have estimated that the ozone layer as we now have, has been formed very slowly over a period of 2500 million years. Several factors—geological, hydrological, biological and lithological aspects of evolution have contributed to the formation of the ozone layers.

The oxygen molecules absorb sunlight (energy) and break up into free oxygen atoms (O)

$$h\vartheta + O_2 \longrightarrow O + O \qquad \ldots (1)$$
$$\text{(energy)}$$

These free individual oxygen atoms combine with oxygen molecules (O_2) to form ozone (O_3).

$$O + O_2 \longrightarrow O_3 \qquad \ldots (2)$$

Ozone is also simultaneously being formed and destroyed by naturally occurring chemical reactions ie, sunlight not only helps in forming ozone, but is also responsible for its breaking up. In this case, an ozone molecule (O_3) is broken into an oxygen atom (O) and an oxygen molecule (O_2).

$$O_3 \xrightarrow{\ h\vartheta\ } O + O_2 \qquad \ldots (3)$$

The free oxygen atom can combine with another ozone molecule to form more oxygen molecules.

$$O + O_3 \longrightarrow O_2 + O_2 \qquad \dots (4)$$

It is interesting to note that even though the ozone layer is about 40 km thick at the high altitudes at which it occurs, in comparison with other atmospheric constituental gases, the abundance of ozone is small. If all the ozone in a vertical column reaching up through the atmosphere were compressed to sea level pressure, the ozone layer would be only a few millimeters thick (Crutzen 1992).

Ozone is found primarily in the stratosphere within the 10 km to 50 km range in altitude which is sometimes referred to as the ozonosphere (Bjerklie 1987).

SIGNIFICANCE OF THE OZONE LAYER

Ozone absorbs most of the harmful ultraviolet rays of the sun, thus preventing them from reaching the earth's surface. The harmful effects of ozone layer depletion on human beings is due to the action of these ultraviolet rays on eyes and skin, causing sun-burns, cataract, skin cancer, etc. The UV rays cause direct damage to the genetic material or DNA of animal cells. Exposure of mammals to UV light has been shown to act on the immune system, thereby making the body more susceptible to diseases.

Thus, it is obvious that the ozone layer acting as a 'sunscreen' prevents the harmful UV rays from reaching the earth's surface directly, thereby helping in reducing the risks of mutation and harm to plant and animal life. This fact underlines the significance of the ozone layer.

THE OZONE HOLE

By 'ozone hole', one does not literally mean a gaping hole in the atmosphere of the earth. It implies that there is a significant decrease in the concentration of ozone in a particular region of the atmosphere. The best example of such an ozone hole is the ozonosphere over the Antarctic which has only about 50 per cent of the ozone that originally occurred there; hence the name 'ozone hole' (Dixit 1992).

Although the ozone layer had been monitored through the 1970s, the actual realization of ozone-depletion came only in 1985. Scientists with the British Antarctic Survey observed that the concentration of ozone in the stratosphere above Antarctica drops rapidly each austral spring, beginning in August, and gradually gets replenished by the end of November (Pitts 1986). The extent and severity of the Antarctic depletion reached a record low in 1993, when the ozone hole extended over more than 9.4 million square miles (Kerr 1993). The World Meteorological Organisation (WMO) has reported that the depletion of ozone in the atmosphere over Antarctica between July and September 1995 was the most rapid ever recorded, and was twice that recorded in '93 and '94.

At first various factors like increased sun-spot activity and the unusual weather systems of the Antarctic were thought to be the cause of the decrease in the ozone concentration. Later it was revealed that anthropogenic sources were also responsible (Grasdel 1978).

The equilibrium between the formation and destruction of ozone, has been upset by the influx of several substances into the atmosphere which react with ozone and destroy it. The rate at which ozone is being destroyed is much faster than the rate at which it is being formed.

Factors accelerating ozone depletion

The main factors responsible for ozone depletion are generally by-products or emissions from industries. The main sources are: *Chlorofluorocarbons (CFCs)*: These molecules are made up of chlorine, fluorine and carbon. Some of the common CFCs and their commercial names are given below (Isaksen 1992).

Dichlorodifluoro methane (Freon 12)
Trichlorofluoro methane (Freon 11)
Chlorodifluoro methane (Freon 22)
Dichlorotetrafluoro ethane (Freon 114)
Trichlorotrifluoro ethane (Freon 113)

Table 3.1 presents a summary of the commonly used CFCs and their applications.

The properties of CFCs like non-corrosiveness, non-inflammability, low toxicity and chemical stability are very useful. Hence, CFCs find a wide and varied application. They are used as refrigerants, propellents in aerosol sprays, foaming agents in plastic manufacturing, fire extinguishing agents, solvents for cleaning electronic and metallic components, for freezing foods etc.

Table 3.1 Commonly used CFCs

Substance (R = Freon)	Chemical Formula	Application	Estimated total world annual production rates, 10^6 kg/ year (1986)
R–11	CCl_3F	Refrigerant, aerosols, foam blowing, solvents, etc.	385
R–12	CCl_2F_2	Refrigerant, aerosols, foam blowing, solvents, etc.	465
R–22	$CHCl_2F$	–	
R–113	CCl_2FCClF_2	Aerosols, foam blowing, solvents, etc.	185
R–114	$CClF_2CClF_2$	Refrigerant, aerosols, foam blowing	15
R–115	$CClF_2CF_2$	Refrigerant	15
R–152a	CH_3CHF_3	–	–
R–500	(R–12/R–152a 73.8/26.2% by weight)	–	–
R–502	(R–22/R–115 48.8/51.2% by weight)	–	–

Unlike other chemicals, CFCs cannot be eliminated from the atmosphere by the usual scavenging processes like photo-dissociation, rain-out and oxidation. In fact, the residence time of CFCs in the atmosphere is estimated to be between 40 and 150 years. During this period, the CFCs move upwards by random diffusion, from the troposphere to the stratosphere.

The CFCs enter the atmosphere by gradual evaporation from their source. CFCs can escape into the atmosphere from a discarded refrigerator. Since the CFCs are thermally stable they can survive in the troposphere. But in the stratosphere, they are exposed to UV radiation. The molecules of CFCs exposed to UV radiation break up, thus freeing chlorine atoms. A free chlorine atom reacts with an ozone molecule (O_3), forming an oxygen molecule (O_2) and a molecule of chlorine monoxide (ClO). The molecules of chlorine monoxide further combine with an atom of oxygen. This reaction results in the formation of an oxygen molecule (O_2) and reformation of the free chlorine atom (Cl).

$$Cl + O_3 \longrightarrow ClO + O_2 \qquad \ldots (5)$$

$$ClO + O \longrightarrow Cl + O_2 \qquad \ldots (6)$$

Net reaction: $O_3 + O \longrightarrow O_2 + O_2 \qquad \ldots (7)$

The depletion of O_3 is catalytic, the element that destroys O_3 being reformed at the end of the cycle (Toon 1991). A single chlorine atom destroys thousands of ozone molecules before encountering reactive nitrogen or hydrogen compounds that eventually return chlorine to its reservoirs (Stolarski 1988).

Nitrogen oxides: The sources of oxides of nitrogen are mainly explosions of thermonuclear weapons in the atmosphere, industrial emissions and agricultural fertilizers. Nitric oxide (NO) catalytically destroys ozone.

$$NO + O_3 \longrightarrow NO_2 + O_2 \qquad \ldots (8)$$

$$NO_2 + O \longrightarrow NO + O_2 \qquad \ldots (9)$$

Nitrous oxide (N_2O) is released from soils through denitrification of nitrates under anaerobic conditions and nitrification of ammonia under aerobic conditions. This N_2O can gradually reach the middle of the stratosphere, where it is photolytically destroyed to yield nitric oxide which in turn destroys ozone (Sethi 1991).

Other substances: Other substances that contribute to ozone depletion include:

Bromine containing compounds called *halons* and *HBFCs*, i.e. hydrobromofluorocarbons (both used in fire extinguishers) and *methyl bromide* (a widely used pesticide). Each bromine atom destroys hundreds of times more ozone molecules than does a chlorine atom.

Carbon tetrachloride (a cheap, highly toxic solvent) and *methyl chloroform* or 1,1,1–trichloroethane (used as a cleaning solvent for clothes and metals, and as a propellant in a wide range of consumer products, such as correction fluid, dry-cleaning sprays, spray adhesives and other aerosols).

Role of polar stratospheric clouds in ozone depletion

There are three types of stratospheric clouds. They are:
a) Nacreous clouds: These clouds extend from 10 to 100 km in length and several kilometres in thickness. They are also called 'mother-of-pearl' clouds due to their glow with a sea-shell like iridescence.
b) The second type of clouds contain nitric acid instead of pure water.
c) The third type of clouds have the same chemical composition as nacreous clouds, but form at a slower rate, which results in a larger cloud with no iridescence.

The chlorine released by the breakdown of CFCs exist initially as pure chlorine or as chlorine monoxide but these two forms react further to form compounds that are stable.

$$Cl + CH_4 \longrightarrow HCl + CH_3 \qquad \ldots (10)$$

$$ClO + NO_2 \longrightarrow ClONO_2 \qquad \ldots (11)$$

The stable compounds HCl and $ClONO_2$ are reservoirs of chlorine, and therefore for chlorine to take part in reactions of any sort, it has to be freed (Solomon et al. 1992).

Susan Solomon and co-workers from NOAA Aeoronomy Laboratory and scientists from Harvard University, suggested in 1986, that a correlation existed between the cycle of ozone depletion and the presence of polar stratospheric clouds (PSCs) ie, the ice particles of the clouds provided substrates for chemical reactions which freed chlorine from its reservoirs. Usually the reaction between HCl and $ClONO_2$ is very slow, but this reaction occurs at a faster rate in the presence of a suitable substrate which is provided by the stratospheric clouds at the poles.

The reaction is

$$HCl + ClONO_2 \longrightarrow Cl_2 + HNO_3 \qquad \ldots (12)$$

It results in the formation of molecular chlorine and nitric acid.

The molecular chlorine formed in the above reaction can be broken down to atomic chlorine and the ozone depletion reaction would continue.

The PSCs not only activate chlorine, but they also absorb reactive nitrogen. If nitrogen oxides were present they would have combined with chlorine monoxide to form a reservoir of chlorine nitrate ($ClONO_2$).

Dimer of chlorine monoxide: It was discovered that the stratospheric chlorine monoxide reacts with itself forming a dimer Cl_2O_2. This dimer is easily dissociated by sunlight, giving rise to free chlorine atoms which can further react to destroy ozone.

Bromine: Bromine, a member of the halogen family can cause depletion of ozone. The reaction is

$$Br + O_3 \longrightarrow BrO + O_2 \qquad \ldots (13)$$

$$BrO + ClO \longrightarrow O_2 + Br + Cl \qquad \ldots (14)$$

Bromine (Br) combines with ozone forming bromine monoxide (BrO) and oxygen (O_2). The BrO further reacts with chlorine monoxide (ClO) to give oxygen (O_2) and free atoms of bromine (Br) and chlorine (Cl). These free atoms can further react with ozone.

Sulphuric acid particles: These particles free chlorine from molecular reservoirs, and convert reactive nitrogen into inert forms thus preventing the formation of chlorine reservoirs. Hofmann et al. suggest that the ozone loss in the lower stratosphere (11–13 km) may be linked to the presence of volcanic aerosols from the eruption of Mount Hudson in Chile.

• THE OZONE DEPLETION THEORIES

Scientists have advanced a number of theories to explain what causes the ozone hole. Several experiments based on scientific expeditions over the Arctic and the Antarctic have been conducted to test these theories. One principal group of theories has suggested that atmospheric motions alone cause the ozone hole. The proponents of these theories hold that the air circulation patterns over the poles may have gradually changed over time so that the upward moving winds might now blow over Antarctica during the spring. These winds, it has been conjectured, would replace ozone-rich stratospheric air with ozone-poor air from the troposphere. Scientists from Ames Research Centre and Leroy National Centre for Atmospheric Research have shown that these hypotheses are incorrect. According to the dynamic models proposed by the advocates of the circulation theories, high concentrations of trace gases originating from the ground should be present at the altitude of the ozone hole. However, investigations showed only low levels of trace gases.

A second class of theories proposes that chemical reactions deplete ozone. One early hypothesis suggested that reactive nitrogen compounds, normally the most important agents for destroying ozone in the lower atmosphere, might exist at elevated concentrations near the ozone hole. The enhancement was presumed to result from the combined effects of increased solar activity and atmospheric circulation. The theory proposes that the enhanced solar activity produces reactive forms of nitrogen over the South Pole at high altitudes. The downward motion of air carries the reactive nitrogen into the lower stratosphere, where investigations reveal a decrease in the concentration of ozone. But Crofton B. Farman and his colleagues from the NASA Jet Propulsion Laboratory and George H. Mount from the National Oceanic and Atmospheric Administration (NOAA) Aeronomy Laboratory found that the reactive forms of nitrogen are also low in the ozone hole, hence disproving this theory.

Farman and his colleagues proposed an alternative chemical interpretation—one that has now gained wide acceptance. Based on the mid-1970s work by Mario J. Molina, of the Massachusetts Institute of Technology, and F. Sherwood Rowland of the University of California at Irvine; the theory suggests that chlorine compounds might be responsible for the ozone hole. Chlorine primarily enters the atmosphere as a component of chlorofluorocarbon (CFC) produced by different industries. Winds throughout the troposphere uniformly distribute CFC moleules released from a single point. After decades, the molecules eventually reach the middle stratosphere. The UV light then tears them apart. Chlorine reservoirs themselves do not destroy the ozone layer. In these compounds, chlorine remains inert and cannot react with the ozone. Early models concluded that CFCs should not have a major effect on the ozone layer. Evidently some mechanism in the Antarctic stratosphere was freeing more of the chlorine from its inert reservoirs. Susan Soloman and her co-workers at the NOAA Aeronomy Laboratory and Michael B. Metlroy and his co-workers at Harvard University attempted to explain the probable mechanism involved. In 1986, they suggested an observed correlation between ozone depletion and the presence of polar stratospheric clouds, implying that chemical reactions on the ice particles in the clouds freed chlorine from the reservoirs. Initially the cloud theory was accepted with hesitation as clouds in the stratosphere were thought to be uncommon. However, further research work confirmed the presence of three types of polar stratospheric clouds (PSCs): nitric acid trihydrate, slowly cooling water-ice and rapidly cooling water-ice clouds (nacreous) which can act as key components in the Antarctic ozone depletion. The PSCs can activate chlorine on their surface as well as use up reactive nitrogen, which would otherwise transfer chlorine to its reservoirs. Slowly cooling water-ice and nitric acid trihydrate clouds can entirely deplete the stratosphere of nitrogen (Toon and Turco 1991). Studies by the UARS (Upper Atmospheric Research Satellite) have also confirmed the presence of fluorine in the stratosphere which corroborates the role of CFCs in ozone depletion. Although the presence of chlorine in the stratosphere can be attributed to volcanic eruptions, salt sprays and the like, apart from the presence of CFCs, the source of fluorine in the stratosphere must be solely attributed to the CFCs.

ENVIRONMENTAL EFFECTS OF
OZONE DEPLETION—AN OVERVIEW

Let us see what are the consequences of ozone depletion during the coming decades, exemplified by the changes in solar ultraviolet radiation reaching the Earth's surface, and the effects on humans, animals, plants, micro-organisms, air quality and materials. The thrust on bio-geochemical cycles which has been neglected for long, has also been taken into account. The main questions now from the point of view of policy are: what will be the most important effects, and what can be done to prevent or mitigate these?

These questions are more difficult to answer than those posed initially when the problem of ozone depletion arose. The question then was, will there be any effects so detrimental as to necessitate protection of the ozone layer? In principle, this could be answered by giving one or two clear-cut examples. The present questions are much broader, and require quantitative knowledge on all effects of potential importance.

The salient features of the assessment made in 1994 by the panel on environmental effects of ozone depletion (Van der Leuen et al. 1995) are discussed below:

A change in the composition of the stratosphere becomes relevant to society only if it has noticeable effects. Hence, the assessment of effects has a major role in the problem of ozone depletion.

Decreases in the quantity of total-column ozone, as now observed in many places, tend to cause increased penetration of solar UV-B radiation (290–315 nm) to the earth's surface. UV-B radiation is the most energetic component of sunlight reaching the earth's surface. It has profound effects on human health, animals, plants, micro-organisms, materials and on air quality. Thus, any perturbation which leads to an increase in UV-B radiation demands careful consideration of the possible consequences.

Changes in ultraviolet radiation

The accuracy of UV measurement has improved greatly of late owing to the sophisticated instrumental techniques employed. Long-term trend detection is still a problem, with little historical data available for baseline estimation.

The amount of UV radiation filtering through the atmosphere to the earth increases with the thinning of the ozone layer. Measurements show that maximum UV levels at the South Pole are reached well before the summer solstice, and DNA-damaging radiation at Palmer Station, Antarctica (64°S) during the springtime ozone depletion can exceed maximum summer values at San Diego, USA (32°N). UV increases at mid-latitudes are smaller. However, increases associated with the record low ozone column of 1992–93 in the northern hemisphere are evident when examined on a wavelength-specific basis (Appenzeller 1993).

Measurements in Argentina, Chile, New Zealand and Australia, show relatively high UV levels compared to corresponding northern hemisphere latitudes, with differences in both stratospheric ozone and tropospheric pollutants likely to play a role. Tropospheric ozone and aerosols can reduce global UV-B irradiance appreciably. At some locations, tropospheric pollution may have increased since pre-industrial times, leading to a decrease in surface UV radiation. However, recent trends in tropospheric pollution probably had only minor effects on UV trends relative to the effect of stratospheric ozone reductions.

Global ozone measurements from satellites over 1979–93 show significant UV-B increases at high and mid-latitudes of both hemispheres, but only small changes in the tropics. Such estimates however assume that the cloud cover and tropospheric pollution have remained constant over these years. Under the current CFC phase-out schedules, global UV levels are predicted to peak around the turn of the century in association with peak loading of chlorine in the stratosphere and the concomitant ozone reductions. The recovery to pre-ozone depletion levels is expected to take place gradually over the next 50 years.

Effects on human and animal health

The increase in UV-B radiation associated with stratospheric ozone depletion is likely to have a substantial impact on human health. Potential risks include an increase in the incidence of and morbidity from eye diseases, skin cancer and infectious diseases. Quantitative estimates of risk are available for some effects (e.g., skin cancer), but others (e.g., infectious diseases) are currently associated with pollution with considerable uncertainty.

UV radiation has been shown in experimental systems to damage the cornea and lens of the eye. Chronic exposure to UV-B (a high, cumulative, lifetime dose) is one of the several factors clearly associated with the risk of cataract of the cortical and posterior subcapsular forms.

Some components of the immune system are present in the skin, which makes the immune system accessible to UV radiation. Experiments in animals show that UV exposure decreases the immune response to skin cancers, infectious agents and other antigens and can lead to unresponsiveness upon repeated challenges. Studies on human subjects also indicate that exposure to UV-B radiation can suppress the induction of some immune responses. The importance of these immune effects for infectious diseases in humans is unknown. However, in parts of the world where infectious diseases already pose a significant challenge to human health, and in persons with impaired immune function, the added impact of UV-B induced immune suppression could be significant.

In susceptible (light-skin coloured) populations, UV-B radiation is the key risk factor for development of non-melanoma skin cancer (NMSC). Using information derived from animal experiments and human epidemiology, it has been estimated that a sustained 1 per cent decrease in stratospheric ozone will result in an increase in NMSC incidence of approximately 2 per cent. The relationship between UV-B exposure and melanoma skin cancer is less well understood and appears to differ fundamentally from that of NMSC. Epidemiologic data indicate that the risk of melanoma increases with exposure to sunlight, especially during childhood. There is, however uncertainty about the relative importance of UV-B radiation, which directly determines the magnitude of the increase in melanoma that would result from ozone depletion.

Effects on terrestrial plants

Physiological and developmental processes of plants are affected by UV-B radiation, even by the amount of UV-B in present-day sunlight. Plants also have several mechanisms to ameliorate or repair these effects and may acclimate to a certain extent to increased levels of UV-B. Nevertheless, plant growth can be directly affected by UV-B radiation.

Response to UV-B also varies considerably among species and also cultivars of the same species. In agriculture, this will necessitate using more UV-B tolerant cultivars and breeding new ones. In forests and grasslands, this is likely to result in changes in the composition of species; therefore there are implications for the biodiversity in different ecosystems.

Indirect changes caused by UV-B (such as changes in plant form, biomass allocation to parts of the plant, timing of developmental phases and secondary metabolism) may be equally or sometimes more important than the damaging effects of UV-B. These changes can have important implications for competitive balance of plants, herbivory, plant pathogens and bio-geochemical cycles. These effects at the ecosystem-level can be anticipated, but not easily predicted or evaluated. Research at the ecosystem level for solar UV-B is barely beginning. Other factors including those involved in climate change such as increasing CO_2 also interact with UV-B. Such reactions are not easily predicted, but are of obvious importance in both agriculture and in non-agricultural ecosystems.

Effects on aquatic ecosystems

More than 30 per cent of the world's animal protein for human consumption comes from the sea, and in many countries, particularly the developing countries, this percentage is significantly higher. As a result, it is important to know how increased levels of exposure to solar UV-B radiation might affect the productivity of aquatic systems.

In addition, the oceans play a key role with respect to global warming. Marine phytoplankton are a major sink for atmospheric carbon dioxide, and they have a decisive role in the development of future trends of CO_2 concentrations in the atmosphere.

Phytoplankton form the foundation of aquatic food webs. Marine phytoplankton are not uniformly distributed throughout the oceans of the world. The highest concentrations are found at high latitudes while, with the exception of upwelling areas on the continental shelves, the tropics and subtropics have 10 to 100 times lower concentrations. In addition to nutrients, temperature, salinity and light availability; the high levels of exposure to solar UV-B radiation that normally occur within the tropics and subtropics may play a role in phytoplankton distributions.

Phytoplankton productivity is limited to the euphotic zone, the upper layer of the water column in which there is sufficient sunlight to support net productivity. The position of the organisms in the euphotic zone is influenced by the action of wind and waves. In addition, many phytoplankton are capable of active movements that enhance their productivity, and therefore, their survival. Exposure to solar UV-B radiation has been shown to affect both orientation mechanisms and motility in phytoplankton, resulting in reduced survival rates for these organisms.

Researchers have measured the increase in, and penetration of UV-B radiation in Antarctic waters, and have provided conclusive evidence of direct ozone-related effects within natural phytoplankton communities. Making use of the space and time variability of the UV-B front associated with the Antarctic ozone hole, researchers assessed phytoplankton productivity within areas under the hole compared to that in areas outside the hole. The results show a direct reduction in phytoplankton productivity due to ozone-related increases in UV-B. One study has indicated a 6–12 per cent reduction in phytoplankton productivity in the marginal ice zone.

Solar UV-B radiation has been found to cause damage in the early developmental stages of fish, shrimp, crab, amphibians and other animals. The most severe effects are decreased reproductive capacity and impaired larval development. Even at current levels, solar UV-B radiation is a limiting factor, and small increases in UV-B exposure could result in a significant reduction in the size of the population of organisms commonly consumed by humans.

Although there is overwhelming evidence that increased UV-B exposure is harmful to aquatic ecosystems; the potential damage can at the present only be roughly estimated.

Effects on bio-geochemical cycles

Increases in solar UV radiation could affect terrestrial and aquatic bio-geochemical cycles, thus, altering both sources and sinks of greenhouse and chemically important trace gases, e.g. carbon dioxide (CO_2), carbon monoxide (CO), carbonyl sulphide (COS) and possibly other gases, including ozone. These potential changes would contribute to biosphere-atmosphere feedbacks that attenuate or reinforce the atmospheric build-up of these gases.

In terrestrial ecosystems, increased UV-B could modify both the production and decomposition of plant matter with concomitant changes in the uptake and release of atmospherically important trace gases. Decomposition processes can be accelerated when UV-B photo-degrades surface litter, and retarded when the dominant effect is on the chemical composition of living tissues, resulting in reduced bio-degradability of buried litter. Primary production can be reduced by enhanced UV-B levels, but the effect is variable between species and even cultivars of some crops. Likewise, photo-production of CO from plant matter is species-dependent and occurs more efficiently from dead than living matter.

In aquatic ecosystems, solar UV-B radiation might also have significant impacts. Studies at several locations have shown that reductions in current levels of solar UV-B result in enhanced primary production, and Antarctic experiments under the ozone hole demonstrated that primary production is inhibited by enhanced UV-B. In addition to its effects on primary production, solar UV radiation can reduce bacterioplankton growth in the upper ocean layer with potentially important effects on marine bio-geochemical cycles. Solar UV radiation stimulates the degradation of aquatic dissolved organic matter (DOM) resulting in loss of UV absorption and formation of dissolved inorganic carbon (DIC), CO, and organic substrates that are readily mineralized or taken up by aquatic micro-organisms. Aquatic nitrogen cycling can be affected by enhanced UV-B through inhibition of nitrifying bacteria and photodecomposition of simple inorganic species such as nitrate. The marine sulphur cycle may be affected by UV-B radiation resulting in possible changes in the sea-to-air emissions of COS and dimethylsulphide (DMS), two gases that are degraded to surface aerosols in the stratosphere and troposphere, respectively.

Recent research on the environmental fate and impact of the hydrofluorocarbon (HFC) and hydrochlorofluorocarbon (HCFC) substitutes for CFCs has focused on trifluoroacetate (TFA), a tropospheric oxidation product of certain HFCs and HCFCs. TFA is mildly toxic to most marine and freshwater phytoplankton. The results indicate that TFA, which may become globally distributed with the increased usage of alternative fluorocarbons, is not likely to accumulate in soils and organisms. Although resistant to chemical degradation, recent evidence indicates that TFA can be broken down by micro-organisms.

Effects on air quality

Reduction in stratospheric ozone and the concomitant increase in UV-B radiation penetrating to the lower atmosphere result in higher photodissociation rates of key trace gases that control the chemical reactivity of the troposphere. This can increase both production and destruction of ozone (O_3) and related oxidants such as hydrogen peroxide (H_2O_2), which are known to have adverse effects on human health, terrestrial plants, and outdoor materials. Changes in the atmospheric concentrations of the hydroxyl radical (OH) may change the atmospheric lifetimes of climatically important gases such as methane (CH_4) and the CFC substitutes.

Trends in the photodissociation rate coefficient of tropospheric O_3, about $+ 0.36 \pm 0.04$ per cent per year in the northern hemisphere and $+ 0.40 \pm 0.05$ per cent per year in the southern hemisphere, have been estimated from satellite measurements of the ozone column between 1979 and 1992. The corresponding model-calculated changes in tropospheric chemical composition are nonlinear and sensitive to the prevailing levels of nitrogen oxides (NO_x). In polluted regions (with high NO_x), tropospheric O_3 is expected to increase, reaching potentially harmful concentrations earlier in the day, and leading to more frequent exceedance of oxidant standards for air quality in urban areas where O_3 levels are routinely near such air-quality thresholds. In more pristine regions (with lower NO_x) O_3 increases can be lower or even negative. Other oxidants such as H_2O_2 and OH are projected to increase in both polluted and pristine regions. Changes in H_2O_2 concentrations may have some impact on the geographical distribution of acid precipitation. Rural regions may become more urban-like and the percentage of areas with remote tropospheric conditions may decline.

Increases in OH concentrations cause a nearly proportionate decrease in the steady-state tropospheric concentrations of CH_4 and CFC substitutes such as the HCFCs and HFCs. Thus, the measured reduction in the ozone column (TOMS 1979–92; Total Ozone Mapping Spectrometer) are likely to have moderated CH_4 increases over the past decade and may account for about one-third of the slowing of the global trend of an increase in CH_4 .

Increased tropospheric reactivity could also lead to increased production of particulates such as cloud condensation nuclei, from the oxidation and subsequent nucleation of sulphur, of both antropogenic and natural origin (e.g., carbonyl sulphide and dimethylsulphide). While these processes are still not fully understood, they exemplify the possibility of complex feedbacks between stratospheric ozone reductions, tropospheric chemistry and climate change.

Effects on materials

Synthetic polymers, naturally occurring biopolymers, as well as some other materials of commercial interest are adversely affected by solar UV radiation. The application of these materials, particularly plastics, in situations which demand routine exposure to sunlight is only possible through the use of light-stabilizers and/or surface treatment to protect them from sunlight. Any increase in solar UV-B content due to partial ozone depletion will therefore accelerate the photogradation rates of these materials, limiting their life outdoors.

The nature and the extent of such damage due to increased UV-B radiation in sunlight is quantified in action spectra. In spite of several polymer action spectra being available in research literature, the information is often inadequate for making reliable estimates of the increased damage. However, it is clear from the available data that the shorter UV-B wavelength processes are mainly responsible for photo-damage ranging from discoloration to loss of mechanical integrity. The molecular level interpretation of these changes remain unclear in many instances.

Higher levels of conventional light-stabilizers in polymer formulations can be used to mitigate the effects of increased UV levels in sunlight. However, such an approach assumes that a) these stabilizers continue to be effective under spectrally altered sunlight conditions; b) they are themselves photostable on exposure to UV-rich sunlight; and c) they can be sufficiently effective at low enough concentrations to serve the purpose. Experimental data bearing on these issues is sparse. Ongoing research, particularly that relating to extreme-environment exposure of polymers is expected to shed more light on these questions. The substitution of the affected materials by more photostable plastics and other materials also remains an attractive possibility. Both these approaches will add to the cost of plastic products in target applications.

Conclusions

The increase in UV-B radiation already observed and expected in the future will have significant consequences in several respects (Van der Leuen et al. 1995). This applies to the UV-B increase predicted on the basis of the most favourable scenario of ozone depletion; it applies even more if the ozone depletion is greater, for instance if there is incomplete compliance with the phase-out agreed for ozone depleting chemicals. Hence, it is clear that strenuous and concerted efforts should be undertaken globally in a sustained manner for the protection of the ozone layer 'umbrella' which wards off harmful UV-B radiation.

THE GLOBAL OZONE DEPLETION TRENDS

Figure 3.3 depicts the Antartic ozone hole. This artist's conception indicates the dramatic reduction discovered in the ozone layers in 1991 compared to the 1989 survey.

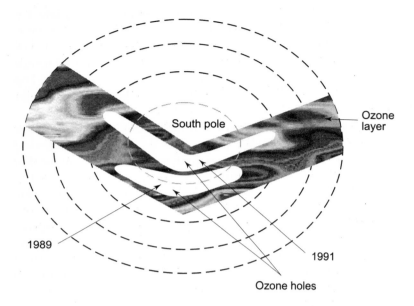

Fig. 3.3 Ozone holes over the Antarctic

The data from the world-wide network of ozone-measuring stations are reported to the World Ozone Data Centre in Toronto and are published monthly.

WHY IS THE HOLE PREDOMINANTLY AT THE ANTARCTIC?

The following factors are believed to be responsible for the preferential depletion of the ozone layer over the Antarctic rather than the Arctic:

1. The Antarctic stratosphere is much colder. The low temperature enables the formation of Polar Stratospheric Clouds (PSCs), below 20 km.

2. Ozone absorbs sunlight, causing the characteristic increase in temperature with increase in altitude in the stratosphere. If ozone is being depleted, the air is cooler, further adding to the favourable conditions for the formation of PSCs and stabilization of the vortex. The vortex is a ring of rapidly circulating air that confines the ozone depletion.

3. The longevity of the Antarctic vortex is another factor, enhancing favourable conditions for the depletion of ozone. The vortex remains, in fact, throughout the polar winter, well into midspring whereas the vortex in the Arctic disintegrates by the time the polar spring (March–April) arrives.

Typical happenings in the winter months leading to the ozone hole over the Antarcatic:

June: Antarctic winter begins, the vortex develops and the temperature falls enough for the clouds to form.

July and August: PSCs denitrify and dehydrate the stratosphere through precipitation; hydrochloric acid and chlorine nitrate react on cloud surfaces to free chlorine, and winter temperatures drop to their lowest point.

September: Sunlight returns to the centre of the vortex as the austral spring begins and PSCs disapper because of increasing temperature; ClO-ClO and ClO-BrO catalytic cycles destroy ozone.

October: Lowest levels of ozone are reached.

November: Polar vortex breaks down, ozone-rich air from the mid-latitudes replenishes the Antarctic stratosphere and ozone-poor air spreads over the southern hemisphere.

Arctic ozone hole

Of late, the ozone hole has been increasingly evident over the Arctic as well. The Arctic ozone hole which swept across Britain in March '96 was the deepest ever seen in the northern hemisphere. Scientists claim that it had been caused, in part, by a dramatic cooling of the upper atmosphere in the northern latitudes over the past two years.

The ozone depletion over the northern hemisphere has been increasing steadily since the winter of 1992. Apart from the build-up of ozone depleting chemicals, the main cause is the increasingly cold temperature in the Arctic stratosphere which encourages the formation of PSCs. Data on temperature in the upper atmosphere collected by weather forecasters and analysed by Steve Pausen at Free University in Berlin show that at an altitude of 20 km, the winter air has been colder during the 1990s than at any time since recording began.

OZONE DEPLETION CONTROL STRATEGIES

The Vienna Convention for the protection of the ozone layer was established in 1985, under the United Nations Environment Programme (UNEP). The convention was held for the promotion of exchange of information, research and systematic observations to protect the human environment and health.

The Montreal Protocol: It was signed by 40 countries in September 1987. An agreement was reached between the manufacturers and consumers on the issue of CFCs. The CFCs were classified under two groups, I and II. The consumption and production of the CFCs in the group–II category were to be stopped at once.

In the case of CFCs in group I, it was agreed upon to implement

- a freeze on consumption at 1986 levels by 1990
- a cut to 80 per cent by 1994
- a further cut to 50 per cent by 1999
- production could be raised by 10 per cent till 1990 but production had to be reduced to 90 per cent by 1994 and to 65 per cent by 1999.

Although reluctant initially, India and China, two potentially important CFC consumers, have finally signed the Montreal Treaty. In the Indian context, the Montreal Protocol seeks to phase out production of CFCs in developing countries by the year 2010. The cost of switching over to CFC substitutes is estimated to be between Rs 35,000 and Rs 60,000 million, depending on the time taken for the change over.

In June 1990, 93 countries signed a United Nations agreement to stop producing CFCs by the end of the 1990s. The main reason for this agreement was the establishment of a US $ 240 million buffer fund to help developing countries develop alternatives to CFCs.

SUBSTITUTES OF CFCs

The continued growth and development of refrigeration and air-conditioning industry in the 1990s and beyond are threatened by unprecedented challenges of change. Escalating energy crisis, global economic concerns and environmental implications have accelerated the eclipse of fully halogenated CFCs which are the basic vapour-compression-cycle refrigerants in vogue today. The recent scientific findings have confirmed CFCs potential destroying the stratospheric ozone layer. Hence the need for substitution.

CFC substitutes—characteristics

The substitutes for CFCs should be safe, effective refrigerants with low ozone layer depletion potential (ODP) and low global warming potential (GWP). The important characteristics of CFC substitutes are as follows:

1.	Chemical properties	– Stable and inert.
2.	Health, safety and environmental considerations	– Non-toxic, non flammable, low ODP, low GWP, non polluting.
3.	Thermophysical properties	– Suitable critical point and boiling point, low molecular and vapour heat capacity, low viscosity and high thermal conductivity.
4.	Miscellaneous	– Soluble in lubricating oil, high vapour-dielectric strength, easy detectibility, low cost and compatibility with refrigeration equipment and materials.

No compound exists which can be used as a substitute-refrigerant for any product category without varying degrees of hardware modification. The key problem areas in the quest for CFC substitutes are the cost of CFC alternatives and the associated replacement technologies, energy efficiency of CFC replacement technology, and the global warming potential. Around 60 per cent of all CFC usage is energy related. Two-thirds of this is used as refrigerants while one-third is used as blowing agents in foam insulation products.

CFC-12 (R-12) is a widely used refrigerant. HFC 134a (R-134a) is the most promising alternative, R-143a and R-152a can also be used.

HCFC-123 (R-123) is the strongest contender to replace CFC-11 (R-11) today. HCFC-123 has an ozone depletion factor of 0.02 and most of it breaks down before reaching the upper atmosphere and the ozone layer. Compatibility of R-123 with

polymeric materials is an important issue that is receiving attention. Table 3.2 presents environmental attributes of CFCs and their
potential substitutes. In general, alternative refrigerants have different performance characteristics than the CFCs they are tar-
geted at replacing. The impact of CFC phase-out on major refrigeration and air-conditioning product categories is shown in table 3.3. For entirely environmentally-benign technology, solar/
waste heat-augmented, vapour-absorption refrigeration systems
should be developed to make them commercially viable. Solar
refrigeration and air-conditioning can be another effective alternative besides the hunt for safer CFC alternatives.

As machines using the conventional refrigerants are
converted to those employing alternative refrigerants, recovery
and recycling of the CFCs presently in use, is also feasible
(Gajendragadkar 1996). Equipment retro-fitting to use the new
refrigerant is the best alternative for chillers. Recovery, recycling
and reclaiming are the compulsory consequences of converting
refrigeration systems containing CFCs to other less reacting
refrigerants (Hoffman 1990).

MONITORING THE OZONE LAYER

Some organizations that help in monitoring the atmosphere and
form a network of information communication about the atmosphere, including ozone layer monitoring are:

World Meteorological Organization (WMO)
World Weather Watch (WWW)
Integrated Global Ocean Services Systems (IGOSS)
Global Ozone Observing System (GO$_3$OS)
Background Air Pollution Monitoring Network (GAW)
Global Climate Observing System (GCOS)

The ozone measurement instruments and techniques are
varied. Some of them are the Dobson spectrophotometer and
the filter ozonometer called M83, and total ozone mapping
spectrometer (TOMS) in the Nimbus-7 satellite.

The Umheher technique: The most common measure of total ozone
abundance is the *Dobson unit*—named after the pioneering
atmospheric physicist Gordon Dobson, which is the thickness

Table 3.2 Environmental attributes of CFCs and substitutes

Substance	Global warming potential (a)	Ozone depletion potential (b)	Commercial status	Status of toxicological testing (Feb 90)	Flam-mability
CFC-11	1500	1.0	in production	tested	none
CFC-12	4500	1.0	"	"	"
HFC-152a	47	0	"	"	mild
HCFC-123	29	0.013–0.027	in pilot production	being tested	none
HFC-32	(c)	0	was produced in limited quantites	limited testing	very mild
HCFC-124	150	0.013–0.030		being tested	none
HCFC-22	510	0.032–0.071	in production	tested	essentially none
HFC-134		0	in pilot production		none
HFC-134a	420	0	"	being tested	"
HCFC-142b	540	0.035–0.77	in production	tested	mild
HCFC-141b	150	0.065–0.14	in production	being tested	"
HFC-125	860	0		prelimi-nary testing	none
HFC-143	1000	0		"	"

Table 3.3 Impact of CFC phase out

Product category	Refrigerant		Impact
	From	To	
Domestic appliances	R-12	R-13a, R-22, blends, mixtures R-290, R-717	major
Residential air-conditioner	R-22	R-22	nil
Automotive air-conditioner	R-12	R-134a, blends, R-22	major
Chillers (positive displacement)	R-22	R-22	nil

Contd.

Table 3.3 (Contd.)

Product category	Refrigerant		Impact
	From	To	
Chillers (centrifugal)	R-11(80%) R-12(10%) R-114	R-123, R-134a, blends, R-124	major
Heat pumps (non-industrial)	R-22	R-22	nil
Heat pumps (industrial, including food storage)	R-22, R-717 R-12 R-502	R-22, R-717	minor
Commercial air conditioning and refrigeration appliances (stores and supermarkets)	R-502, R-22 R-12	R-22, R-125	minor

of the ozone column (compressed to STP) in milli-centimeters. At STP (Standard Temperature and Pressure), one Dobson unit is equal to 2.69×10^{20} molecules per square meter.

Two major changes observed in atmospheric ozone in the last few years are: (a) the strong downward trend of spring time Antarctic ozone since 1980, and (b) the widespread decrease in total ozone in the subtropical and middle latitudes since 1982, associated with the eruptions of El Chichon in March-April 1982, which has recurred in succeeding years. The likely causes are either influences of atmospheric chemistry such as nuclear tests, volcanic eruption, injection of CFCs, etc, or changes in atmospheric large scale dynamics (Mani A. 1993).

Although the overall ozone levels in the atmosphere are routinely measured throughout the world by spectrographic Dobson instruments on the ground and by the TOMS, the subtle and continuously changing structural details of the atmosphere that influence the rate and degree of ozone destruction remain obscure.

For this reason, Dr Chester S. Gardner, Dr George C. Papen and their colleagues from the University of Illinois have installed a year-round laser radar, or 'lidar', designed to maintain sur-veillance of the atmosphere above the South Pole at altitudes up to 60 miles. The instrument allows scientists to create continuous three-dimensional maps of the chemical composition and physical details of the atmosphere throughout the year.

CONCLUSION

At the very basis of modern civilization and comforts are the substances which can destroy ozone. Even if all production was curtailed immediately, depletion of ozone would continue to occur for at least a decade. This would be partly due to the slow movement of already emitted CFCs into the photolytically active middle stratosphere. If management strategies are not adopted, the Antarctic along with the Arctic and rest of the northern hemisphere would be seriously exposed to the consequences of stratospheric ozone depletion. But all has not been lost yet; there is hope.

A major impetus to research efforts on the ozone hole has recently come in the form of the 1995 Nobel prize for chemistry to three environmentalists for their contributions towards research on the ozone hole: Paul Crutzen, Mario Molina and F. Sherwood Rowland.

Beginning in the 1970s, the trio, in a series of findings described the chemical processes by which ozone was formed and destroyed in the atmosphere. Their research threw light on the role of NO_x in accelerating ozone depletion, the link between the density of ozone layer and chemicals released by the bacteria in the soil, and the threat to the ozone layer by the proliferation of the synthetic CFCs. The work of Molina, Crutzen and Rowland laid the foundation for the Montreal Protocol in 1987, on phasing out of CFCs by AD 2000. The trail blazed by these great scientists may well serve as the first step towards greater understanding of the enigmatic 'ozone hole'.

Another heartening fact which has emerged recently is that, for the first time, scientists have seen a reduction in ozone-eating chemicals in the lower atmosphere. They predict that within four years, this trend will be repeated in the stratosphere, where the real damage to ozone is being done (Pearce 1996). These results indicate that the Montreal Protocol, which limits the production of ozone-depleting chemicals such as chlorofluorocarbons has had some desirable impact. If sustained efforts are made at the global level to contain the ozone hole and prevent it from getting bigger and deeper, it is may be possible to achieve success and even reverse the trend.

REFERENCES

Appenzeller, T. (1993) 'Filling a Hole in the Ozone Argument'. *Science* 262.

Bjerklie, D. (1987) 'The Heat is On'. *Time*, 19 October, New York.

Crutzen, P.J. (1992) 'Ultraviolet on the Increase'. *Nature* 356.

Dixit, D.K. (1992) 'Ozone Safe Refrigerants'. *Technorama*, July–September.

Ehhalt, D.H. (1988) 'Greenhouse Gases—Emissions and Sinks'. In *The Changing Atmosphere: Report of the Dahlem Workshop on the Changing Atmosphere,* edited by F.S. Rowland and I.S.A. Isaken. New York: Wiley Interscience, pp. 25–32.

'Focus on Ozone Depletion'. *Deccan Herald*, 1 June, 1992.

Gajendragadkar, S.K. (1996) 'CFCs—Problem and Solutions—A Critical Review'. *Journal of the Indian Association for Environmental Management* 23: 45–9.

Grasdel, T.E. (1978) *Chemical Compounds in the Atmosphere.* New York: Academic Press.

Hoffman, J.S. (1990) 'Replacing CFCs: The Search for Alternatives'. *Ambio* 19.

Hoffman, J.S., J.B. Wells and J.G. Titus (1986) 'Future Global Warming and Sea Level Rise'. In *Iceland Coastal and River Symposium,* edited by G. Sigbjaranson Reykjavik. National Energy Authority, pp. 245–66.

Husain, A., P.C. Joshi and P.K. Ray (1989) 'Ozone Depletion Means Dangers'. *Science Reporter* 26.

Isaksen, I.S.A., V. Ramaswamy, H. Rodhe and T.M.L. Wigley (1992) 'Radiative Forcing of Climate'. In *Climate Change, 1992* (The Supplementary Report of the IPCC Scientific Assessment), edited by J.T. Houghton, B.A. Calendar and S.K. Varacy. Cambridge: Cambridge University Press, pp. 47–67.

Kerr, R.A. (1992) 'Not Over the Arctic—For Now'. *Science* 256.

—— (1993) 'The Ozone Hole Reaches a New Low'. *Science* 256.

Mani, A. (1993) 'Ozone in the Tropics'. *Current Science* 335.

Pearce, F. (1996) 'Big Freeze Digs a Deeper Hole in Ozone Layer'. *New Scientist* 149.

—— (1996) 'Winning of War on Ozone Eaters'. *New Scientist* 150.

Pitts, F.B.J. and J.N. Pitts (1986) *Atmospheric Chemistry Fundamentals and Experimental Techniques*. New York: Wiley.

Sethi, M.S. and I.K. Sethi (1991) *Understanding Our Environment*. New Delhi: Commonwealth Publishers, pp. 119–28.

Solomon, S. and L.D. Albriton (1992) 'Time Dependent Ozone Depletion Potentials for Short and Long Term Forecasts'. *Nature* 357.

Stolarski, S.R. (1988) 'The Antarctic Ozone Hole'. *Scientific American* 258.

——, R. Bojkov, L. Bishop, C. Zerefos, J. Staehelin and J. Zawodny (1992) 'Measured Trend in Stratospheric Ozone'. *Science* 256.

Thompson, A.M. (1992) 'The Oxidizing Capacity of the Earth's Atmosphere: Probable Past and Future Changes'. *Science* 256.

Toon, O.B. and P.R. Turco (1991) 'Polar Stratospheric Clouds and Ozone Depletion', *Scientific American* 264.

Van der Leun, J., Xiaoyan Tang and M. Tevini (1995) 'Environmental Effects of Ozone Depletion 1994 Assessment'. *Ambio* 24.

Vohra, K.G. (1990) 'The Earth Loses its Protective Ozone Screen'. *Science Today* 24.

Hazardous Wastes and Their Management—I
Definition, Characterization, Treatment and Disposal

As more and more new products and processes are developed, new chemicals produced, and materials created, more and more such waste is generated which could be *hazardous*. The term *hazardous waste* was initially used to differentiate highly toxic or offensive wastes from the familiar wastes such as sewage and household garbage. With the passage of time and continual increase in the quantities and *types* of toxic wastes, it became necessary to define the term.

More importantly it became necessary to develop ways and means of analysing, handling and treating hazardous wastes.

HAZARDOUS WASTES

Definition

The task of realistically defining hazardous waste is extremely complex, presenting the first stumbling block in hazardous waste control programmes. As per the definition of the Resource Conservation and Recovery Act (RCRA) of USA, *hazardous waste is a solid, liquid or gaseous waste, or combination of wastes, that because of its quantity, concentration or characteristics may cause or significantly contribute to an increase in mortality or an increase in serious irreversible or incapacitating reversible illness and pose a substantial present or potential hazard to human health or the environment when improperly treated, stored, transported, disposed of or otherwise managed.*

In short a *hazardous waste* is one which is potentially harmful to the eco-system unless properly managed.

The US Environmental Protection Agency (EPA) has defined a waste to be hazardous under the legislation if it:

i) exhibits characteristics of ignitability, corrosivity, reactivity and/or toxicity;

ii) is a nonspecific source waste (generic waste from industrial processes);

iii) is a specific source waste (from specific industries);

iv) is a specific commercial chemical product or intermediate;

v) is a mixture containing a listed hazardous waste; or,

vi) is a substance that is not excluded from regulation under RCRA, subtitle C (Wentz 1989).

Hazardous materials are those which are toxic, persistent, ignitable, corrosive, bio-accumulative, infectious or pathogenic.

Examples include wastes from plastic, pesticide, herbicide, medicine, paint and petroleum industries.

Impact of hazardous wastes on the ecosystem—acute and chronic

A few classic cases in point highlighting the danger posed by hazardous wastes are enumerated below:

Love Canal: An area of about 16 acres in Niagra Falls was used from 1942 to 1953 as a dump-site for approximately 22,000 tonnes of chemicals (*Health* 1983). The area was capped with clay and later houses and a school were built up to the edge of the site.

In 1978–79 over 200 families were evacuated as potential teratogenic, mutagenic and oncogenic chemicals were identified from the area leading to increased abortions and birth defects (Holden 1980).

Times Beach: In the winter of 1982–83, over 2,200 people were evacuated from Times Beach, Missouri, owing to the presence of toxic wastes in the soil and water. The toxins belonged to Dioxin (TCD) group, a class of over 75 chemicals, known for their hazardous nature. They are unwanted by-products of organic compounds—chlorinated phenols, widely used in the manufacture of plastic, herbicides and pesticides.

Another group of hazardous wastes found there were PCBs (Polychlorinated Biphenyls) used as coolant liquids in the electrical industry. PCBs are known to damage skin, eyes and lungs, and cause birth defects and cancer.

Hardeman County: In Tennesse, 60 miles north of Memphis, is a 200-acre hazardous waste dump, which was operational between 1964 and 1972. Up to 25 million gallons of solid and liquid wastes were deposited, which eventually contaminated surface and ground water with chlorinated organic compounds and other chemicals (Meyer 1983).

Delhi: Nearer home, a noxious nightmare assailed the residents of Kardampuri colony in East Delhi in the wee hours of 13 November 1994. Toxic fumes from a heap set aflame by a local junk dealer created the panicky exodus. The people complained of severe breathing distress, irritation and pain in the throat, vomiting, and dizziness. Many people, including infants, were seriously affected and a few succumbed to the poisoning. Tests revealed that the chemicals responsible were cyanide, cadmium, selenium and arsenic. Traces of lead, aluminium and copper were also found. It is widely suspected that the burning of metals like cobalt and manganese along with pesticides like organophosphates and carbonates, caused the noxious fumes.

Gujarat: The catalogue of disasters will be incomplete without the inclusion of the recent Surat plague epidemic. The bubonic plague epidemic, which descended on the city of Surat in 1994, owed its origin to the woefully inadequate waste-management strategies. The waste heaps which were allowed to rot resulted in the proliferation of rodents and vermin, which ultimately triggered off the pestilence of plague which had been thought of as having been eradicated long back. As the people were caught unaware, the disease wrought wide-spread havoc. This episode illustrates the fact that at times even those solid wastes which do not come under the definition of 'hazardous wastes', can become extremely hazardous!

HEALTH HAZARDS

The evaluation of the health effects of hazardous wastes depends upon three basic problems:

    i) availability of toxicologic data on chemicals;

ii) toxicity of mixtures of chemical wastes and the physical, chemical and biological factors that influence toxicity; and
iii) estimation of the level and duration of exposure to the population.

The toxic effects induced by hazardous wastes can be differentiated into two different groups: responses that result from genetic effects and those associated with target organs.

In addition to epidemiological methods used in assessing health effects, several basic issues need to be reconciled prior to data interpretation. These include latency period, multiple causative factors, population diversity and mobility, and socio-economic and urbanization conditions.

If the expected baseline frequency of specific health effects is low, a large population base will be needed, especially for risk assessment.

Secondary (health) hazards

The wastes in the natural environmental conditions can become toxic, explosive or even become the breeding ground for disease germs and vectors.

For example, cooked meat can become a lethal source of pathogens within a few hours. Vectors like house flies, mosquitoes, rats etc are indicators of conversion of normal wastes into hazardous ones.

When organic materials are amassed even less flammable materials undergo spontaneous combustion, eg, hay, saw dust, coal, etc.

Anaerobic decomposition of organic wastes produces methane (CH_4), hydrogen sulphide (H_2S), carbon monoxide(CO), carbon dioxide (CO_2), etc. They may form explosive mixtures. The recent explosion due to leakage of gas accumulated in a choked sewage in Mexico is note worthy.

Leachate from even normal municipality wastes can assume gigantic proportions. The Bio-chemical Oxygen Demand (BOD) of leachates from landfill may exceed 20,000 mg/litre, i.e. 100 times stronger than raw wage.

Thus, secondary hazards of normally safe wastes are also studied as a special case of hazardous wastes.

HAZARDOUS WASTE GENERATION

A review of the available literature points to four major pathways for hazardous waste generation and their escape in to the environment. The first and most significant pathway is the continuous discharge of hazardous wastes onto land which include:

- wastes which, due to enviornmental restrictions, have been removed from air or water effluent streams;
- wastes generated by industry as losses or by-products of various processes;
- products which become wastes as a result of governmental regulations or restrictions;
- wastes from government or other institutional operations.

The second source is the accidental spilling of hazardous substances resulting in damage to the environment. Another source is the discharge of hazardous wastes in small quantities leading to its concentration, from numerous sources including waste effluent and air emission streams that are not subject to pollution control procedures. The fourth pathway consists of mine tailings and abandoned dumping and disposal sites contributing to slow enviornmental degradation. The problem is looming large as its deleterious effects are slowly being manifested globally.

The international scenario

In 1984, the US alone produced over 260 million metric tonnes of hazardous wastes equivalent to over 70 billion US gallons (Thomas 1984). There were about 14,000 installations that generate waste and the majority (96%) managed them onsite. The industrial sector produced 92 per cent of all hazardous material with 68 per cent of the total from chemical industries alone. 99 per cent of the material is liquid at the time of generation. The machinery and transportation industries, each contribute about 6 per cent. By 1986 USA had over 22,000 hazardous waste sites. More than 500 have been identified as posing serious long-term hazards to public health and the environment (Thomas, 1984).

Accurate determinations of the amount of waste produced

by other countries are generally not available. In the Netherlands, there are 3000 known abandoned dumps, of which 300 pose an imminent threat to human health (Hinrichsen 1983). In Great Britain, some counties have been hesitant to accept wastes whereas others operate with a more liberal policy (Pearce and Caufield 1981). This country is estimated to have produced 11 million metric tonnes during 1980 (Hinrichsen 1983). France produces about 2–3 million metric tonnes of toxic wastes annually and ships 4000 tonnes of chlorine to be burned at sea. In China, medium to large industries generated approximately 80 million metric tonnes of solid wastes in 1979 (Geping 1982).

HAZARDOUS WASTE CHARACTERIZATION

In order to devise and discuss the techniques for the disposal, handling, storage and recovery of hazardous wastes, it is essential to have a clear idea about the nature of the materials. A complete characterization scheme is necessary for supplying all the relevant details regarding waste materials.

According to the EPA (Environmental Protection Agency of the US) regulations, four characteristics have been identified as inherently hazardous in any substance, depending on the definability in terms of physical, chemical or other properties; and the ease of detection by standardized and available testing protocols. These characteristics are: ignitability, corrosivity, reactivity, and extraction procedure (EP) toxicity.

Ignitability

Ignitability is the characteristic used to declare as 'hazardous', those wastes that could cause a fire during transport, storage or disposal. Examples of ignitable wastes include waste oils and used solvents.

A waste exhibits ignitability if a representative sample of the waste has any one of the following properties:

1. it is liquid, other than an aqueous solution containing less than 24 per cent alcohol by volume, and has a flash point less than 60°C (140°F), as determined using a Pensky-Martens closed cup tester or by a Seta Flash closed cup tester (American Society for Testing and Materials 1987);

2. it is not a liquid and is capable, under standard temperature and pressure, of causing fire through friction, absorption of moisture or spontaneous chemical changes and, when ignited, burns so vigorously and persistently that it creates a hazard;
3. it is an ignitable, compressed gas;
4. it is an oxidizer.

Corrosivity

Corrosivity, mainly measured by pH, is an important characteristic in identifying hazardous wastes, as wastes with high or low pH can react dangerously with other wastes or cause toxic contaminants to migrate from certain wastes. Examples of corrosive wastes include acidic wastes and used pickle liquor from steel manufacture.

A waste exhibits corrosivity if a representative sample of the waste has any one of the following properties:

1. it is aqueous and has a pH less than or equal to 2, or, greater than or equal to 12.5 as determined by the pH-meter using an EPA test method;
2. it is a liquid and it corrodes steel at a rate greater than 6.35 mm (0.250 inch) per year at a test temperature of 55°C (130°F).

Reactivity

Reactivity was chosen as an identifying characteristic of hazardous waste because unstable wastes can pose an explosive problem at any stage of the waste management cycle. Examples include water from TNT operations and used cyanide solvents.

A waste is said to be reactive if a representative sample of the waste has any of the following properties:

1. It is normally unstable and it readily undergoes violent change without detonating;
2. It reacts violently with water;
3. It forms potentially explosive mixtures with water;
4. When mixed with water, it generates toxic gases, vapours or fumes in a quantity sufficient to present a danger to human health or the environment;
5. It is a cyanide or sulphide-bearing waste, which when

exposed to pH between 2.0 and 12.5 can generate toxic gases, vapours or fumes in a quantity sufficient to endanger human health and the environment;

6. It is capable of detonation or explosive decomposition on reaction at standard temperature and pressure;

7. It is capable of detonation or explosive reaction if subjected to a strong igniting source or if heated under confinement;

8. It is a forbidden explosive.

Extraction procedure toxicity

Wastes that exhibit extraction procedure (EP) toxicity are termed as hazardous.

Extraction procedure toxicity is designed to identify wastes that are likely to release hazardous concentrations of a particular toxic constituents into ground water as a result of improper management.

During the procedure, the wastes are leached in a manner that simulates the leaching actions that occur in landfills. The extract is analysed to determine if it possesses certain toxic contaminants. If the concentration exceeds the regulatory levels then the waste is classified as hazardous.

Some of the regulatory levels of the contaminants in the leached extract are given below:

Contaminant	Maximum Concentration (mg/l)
Arsenic	5.0
Barium	100.0
Cadmium	1.0
Chromium	5.0
Lead	5.0
Mercury	0.2
Selenium	1.0
Silver	5.0

Hazardous waste management

Appropriate management techniques have to take into account the various options based on the nature, quantity and location of the waste. The desirable options for managing hazardous wastes are listed below, on priority basis. Figure 4.1 is a schematic representation of the prioritized options.

i) Mimimization of the amount generated by modifying the industrial processes involved.

ii) Transfer of the waste to another industry that can utilize it.

iii) Reprocessing of the waste to recover energy or materials.

iv) Separation of hazardous waste from non-hazardous waste at the source and its subsequent concentration, which reduces the handling, transportation and disposal costs.

v) Incineration of the waste or its treatment to reduce the degree of hazard, and

vi) Disposal of the waste in a secure landfill, one that is located, designed, operated and monitored in a manner that protects life and environment (Wentz 1989). Let us examine the different disposal and treatment methods currently in use.

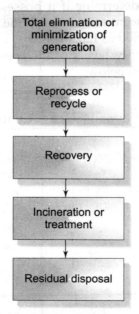

Fig. 4.1 Priorities in hazardous waste management
Source: Wentz 1989

Landfills

Landfills for the disposal of hazardous wastes evolved from sanitary landfills.

SITE SELECTION

The selection of a site for waste disposal is governed by climatic, geologic and hydrological factors. Arid conditions are favourable as little leaching occurs (figure 4.2a). Sites made up of impermeable material, such as claytill, are preferable because such

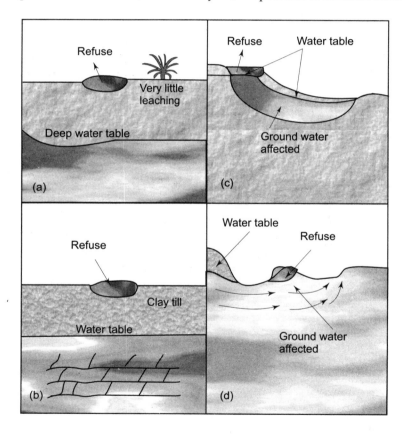

Fig. 4.2 Hydrogeologic conditions for disposal of refuse at land surface:
(a) Dry environment, (b) Low permeability, (c) Recharge zone,
(d) Discharge zone

material retards the movement of contaminants from the site and attenuates the effect by adsorption and filtration (figure 4.2b). On the other hand, a site close to the recharge zone is not suitable because the ground water gets affected (figure 4.2c). Likewise, if the discharge flows into a stream because of a high water table, the ground water gets contaminated (figure 4.2d).

Additional engineering features in landfill design that supplement natural geologic features, increase the efficiency of hazardous waste containment.

DESIGN AND CONSTRUCTION

The design and construction of a hazardous landfill incorporates an impermeable cover, an impermeable bottom liner, a system of drainage pipes to collect and remove any leachate that may accumulate, and a system of monitoring wells (figure 4.3).

The cover prevents infiltration of precipitation into the landfill. The bottom liner serves to reduce the rate of leachate

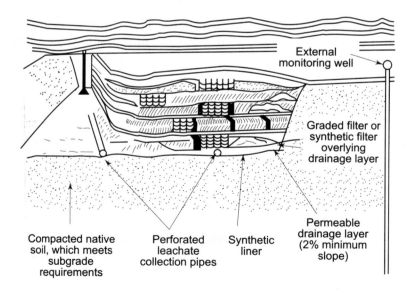

Fig. 4.3 Generalized depiction of a hazardous waste landfill

migration. The leachate collection and recovery system checks leachate migration and hydrostatic pressure. This is accomplished by a system of perforated pipes buried in the lower part of the landfill, from where the liquid is pumped out. Regular monitoring of the wells can detect before hand, the effects of groundwater contamination.

The NAS (National Academy of Sciences) has decided that at least 500 years is realistic as a period of concern for wastes in landfills.

Landfill gas generation is a major concern in organic waste dumpings. The emission rate from landfills depends on a number of factors, such as vapour pressure, diffusion coefficient, mass transfer coefficient and solubility.

Several control techniques have been proposed regarding toxic gas emission from landfills (Shen 1981).

Wastes that are explosive or have a high vapour pressure, such as organic sludges, volatile organic wastes and liquids, should not be landfilled.

Many types of waste should be pre-treated to make them more innocuous and less volatile.

Gas collection devices should be installed.

The site should be capped.

Hazardous wastes are now stored in separate cells, ie, discrete storage areas, which are highly suitable for incompatible wastes. Also, when a cell is full, it can be quickly sealed and revegetated.

The Office of Technology Assessment has concluded that complete protection from migration, even for the operating life of the fill, is probably unattainable.

Land treatment

This is a biological method in which, hazardous wastes are deposited either on the land or injected just beneath it and degraded naturally by aerobic organisms. Oxygen levels may be maintained by periodic ploughing.

Although the basic concept of land treatment is simple, its planning and implementation are exceedingly complex, involving an understanding of microbiology, soil science, chemistry, hydrology, geology and climatology. In order for organic constituents to be suitable for land treatment, they must degrade at a rate faster than volatization, leaching or runoff. The mobility, toxicity and accumulation of heavy metals must also be considered in facility design.

Acceptable sites for landfill may also prove most acceptable for land treatment. If bio-degradable hazardous wastes are eliminated, and toxic metals and other hazardous materials immobilized in surficial material, land treatment could prove to be superior to landfill.

Selection of the kind of hazardous waste treated by this technique, combined with careful site monitoring, will reduce the danger of environmental contamination by the biodegradable fraction of the waste. The ultimate fate of toxic materials that remain in the treated ground is more questionable.

Deep-well injection

The use of injection wells for industrial waste disposal began around 1950. In a typical injection well, depths to the disposal zone commonly range between 600 to 1800 m but may be shallower or deeper.

The major difference between an injection well and a normal well is the closed annular space between the injection tubing and the inner or long string casing. This space is filled with a fluid under pressure which preserves the casing and tubing, and which is monitored by a pressure gauge at the surface. On completion, the well is plugged to prevent the release of liquids and also any change in reservoir pressure.

Underground injection of hazardous wastes requires a very careful appraisal of factors including subsurface stratigraphy, lithology, subsurface structure, fresh-water geohydrology, extent of disposal zone area, pressure conditions of disposal zone, density, toxicity, chemistry and reactivity of wastes, etc.

An ideal disposal reservoir has a thick, porous and permeable blanket of sandstone underlying an entire basin and confined from above and below by impermeable beds. A waste that is lighter than the interstitial water would be effectively contained in an anticline. In contrast, a fluid denser than the interstitial water would be best contained in a syncline (figure 4.4).

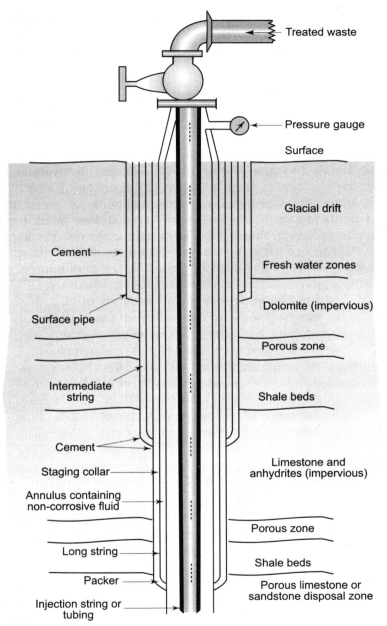

Treated waste

Pressure gauge

Surface

Glacial drift

Cement

Fresh water zones

Dolomite (impervious)

Surface pipe

Porous zone

Intermediate
string

Shale beds

Cement

Limestone and
anhydrites (impervious)

Staging collar

Annulus containing
non-corrosive fluid

Porous zone

Long string

Shale beds

Packer

Porous limestone or
sandstone disposal zone

Injection string or
tubing

Fig. 4.4 Generalized sketch of a disposal well

Another important factor in underground disposal is the selection of an injection pressure high enough to displace the interstitial fluids with the waste but is not so high that the containing impermeable beds are fractured.

Incineration

The safest and most effective alternative to land-based disposal of hazardous wastes is incineration. This process is widely used in Europe, chiefly in Germany.

In 1981, about 240 facilities in the US incinerated 17 million metric tonnes of waste (Sobrino 1984). Another 3–5 million metric tonnes were burned in industrial boliers and 3,50,000 metric tonnes by other means like cement kilns.

Many incinerators have been shut down as they failed to meet air pollution standards. To obtain a permit for an incinerator, the company must demonstrate a removal efficiency of 99.99 per cent for PCBs and 9.99 per cent for other pollutants.

Many organic toxic wastes are broken down to harmless CO_2 and H_2O at high temperatures. Small amounts of HCl, SO_2, dioxins etc, may be produced depending on the efficiency of the incinerator. These can be removed using special equipment.

The effectiveness of incineration depends upon temperature, turbulence and residence time (Josephson 1984). PCBs can be incinerated with a 99.99 per cent efficiency at 750°C. The chief breakdown product, hexachlorobenzene (HCB), decomposes after 800°C and persists at low levels even at 1000°C. Fine particulate matter, metal aerosols and hydrogen halide gases that escape incineration also pose a problem. New methods like ionizing wet scrubbers, supersonic steam injection and electrostatic precipitation will lead to an increased treatment of these wastes.

A new experimental method involves heating a mixture of water and organic wastes to 400°C at high pressure. Air is passed through this mixture. All organic compounds, including dioxins, break into water and CO_2.

High incinerator temperatures result in an increased emission of heavy metals. Gerstly and Albrinck (1982) have shown an increase in vapour pressure of cadmium, copper, lead and chromium with temperature.

Incineration at sea: This is attractive because of its low cost, roughly half that of incineration on land. Interest in sea disposal began in the early 1970s when VOLVANUS I went into operation in the North Sea. In 1977, 10,400 metric tonnes of Agent Orange were burned in the Pacific Ocean near Johnston Atoll. The largest amount to be incinerated was 1.5 million gallons of PCB in the Gulf of Mexico (Bond 1984) in 1982.

To date, incineration at sea can be described as a cheap and effective method.

Other treatment methods

In many instances, these methods are still under development and are not presently cost effective. Heavy metals, if not recycled, poses a potentially long-term toxicologic problem for waste disposal. They can be solidified into granular form, mixed with cement-based grouts and pumped into underground disposal caverns (Forsberg 1984).

Thermosplastic techniques using polyethylene, paraffins and bitumen are popular as the wastes are tightly bound and the leaching rate is low due to the water resistant media.

Several other waste treatment processes that are either under research or in use include:

Physical treatment: Physical treatment of hazardous wastes include a number of separation processes commonly used in industry. For a waste containing liquids and solids, physical separation is of great value as it is simple and very economical. The physical processes for the separation of liquids and solids are:

1. Screening
2. Sedimentation and clarification
3. Centrifugation
4. Floatation
5. Filtration
6. Sorption
7. Evaporation and distillation
8. Reverse osmosis
9. Stripping.

The selection of a particular treatment process depends upon the nature of the constituents and their amenability to the technique.

Biological treatment: This method is now increasingly being used as an efficient, cost-effective way to remove hazardous substances from waste water effluent.

Aerobic organisms (microbes which utilize oxygen directly) are commonly used to treat waste streams and anaerobic organisms (microbes which use oxygen present in chemical combination with other elements) are made use of in the treatment of strong organic wastes or organic sludges.

The ability of bacteria to consume organic matter is measured by the bio-chemical oxygen demand which is the quantity of oxygen utilized by micro-organisms in the aerobic oxidation of organics at 20°C. Hazardous waste materials are toxic to some of the micro-organisms. But a substance that is toxic to one group of organisms may be an essential nutrient for another. By achieving the proper distribution of organisms, biological treatment can be effectively harnassed.

RADIOACTIVE WASTES

Liquid, solid and gaseous wastes are produced in the mining of or in the production of reactor fuel materials, reactor operation, processing of irradiated reactor fuels and numerous other related processes. Wastes also result from the use of radioactive materials in research laboratories, industries and medical treatement.

Based on radioactivity, the radioactive wastes can be classified as

Mildly radioactive: residues from filter and purification processes, contaminated equipment, gloves, sewage sludge from wastewater separation.

Moderately radioactive: component parts of nuclear power stations rendered active by neutrons, radioactive residues from purification plants.

Highly radioactive: includes decaying fissile materials e.g., those of strontiun Sr^{90}, Caesium Cs^{137}, Iodine I^{129}. The half-life periods of these radioactive isotopes are 26, 30, and 17,200,000 years, respectively.

For disposal purposes, nuclear wastes are separated into two groups: *High-level* radioactive wastes (HLRW) which include (Hileman 1982b, White & Spath 1984):

i) spent nuclear fuel after irradiation, fission products and TRU (Trans Uranic);
ii) trans Uranic wastes which are x-emitting TRU isotopes, with half lives of over a year;
iii) high-level wastes (HLW) which are by-products of spent-fuel reprocessing, especially to extract plutonium for warheads.

Low level radioactive wastes (LLRW) include:
i) low level wastes, defined as wastes containing less than 10nCi per gram of trans-uranic elements;
ii) Uranium (U) and Thorium (Th) by-product materials are the tailings produced by the extraction or concentration of U or Th from processed ore.

Until recently, a criterion of 10 $nCig^{-1}$ ($nCig^{-1}$=nanocuries/gm) served as a cutoff between shallow land burial and other modes of disposal for TRU high-level wastes (Moghissi 1984). Proposed standards define concentration limits for specific radionucleides. For x-emiting TRU nuclides with half-life over 20 years, the limit is 100 $nCig^{-1}$. All other radionucleides with a half-life of over 20 years have a maximum of 1 $nCig^{-1}$.

The total volume of LLRW produced in the US varied between 80,000 and 112,000 cubic metres per annum in 1979–82. The corresponding range for radioactivity was 260,000–500,000 Ci. In the US, generation of LLRW is about 10 times greater than that of high-level wastes.

TREATMENT OF RADIOACTIVE WASTES

Approximately one-third and one-fourth of the spent fuel rods in Pressurized water Reactors (PWR) and Boiling water Reactor (BWR) respectively, is removed and replaced. The main objectives of fuel reprocessing are the removal of HLW and TRU from fuel rods and the separation of plutonium. After the initial stages, a nitric solution of the fuel is put in contact with an immiscible solvent, like tributyl phosphate present in an organic diluent. This solution, raffinate, is highly radioactive and is concentrated by evaporation. Raffinate is stored in special stainless steel containers. Approximately 10 cubic metres of concentrated waste is produced from each GW of electricity (Miller 1983).

Reprocessing produces both solid and liquid wastes. Liquids can be solidified by spray calcination and fluidized-bed calcination (Glasstone and Jordan 1980). In fluidized-bed calcination, liquid waste is continuously fed into a calciner containing a bed of small nucleation particles. The bed is heated to 500–600°C by kerosene combustion. A stream of air is passed through the particles so that they flow like a liquid contact between the liquid and particles, causing drying and calcination.

In spray calcination, the liquid waste is sprayed into the top of tower that is heated in the furnace. At about 700°C, water is driven off resulting in calcinated solids, which is collected at the bottom of the tower. Heating them to 900°C drives off the remaining nitrates, whereas, if it is to be vitrified, the powder is heated to 1000–10000°C to form a mass of glass. The 'supercalcine' process produces a calcine with up to 23 per cent additional constituents like lime (Miller 1983).

Radioactive wastes disposal

Storage in tanks above ground: The US has been practising this for over 20 years. There are over 200 steel and concrete tanks having over 3 million litres of highly radioactive liquid. These radioactive wastes generate heat, besides radiation and hence require constant cooling. The heat is transferred to the condenser by pipes carrying steam. Mixing of contents with compressed air ensures uniform heating and does not allow settling of solids.

A few serious loopholes in this method are listed below:

i) Strong radiation from wastes might lead to corrosion of the tanks and a consequent leakage of radioactive wastes, e.g., Hanford, Washington, the prime deposit site in the US experienced seepage of 4,90,000 litres of radioactive waste;

ii) fuming wastes require constant refrigeration. They produce 9 kw/m^3 of highly radioactive waste. In case of a refrigeration failure, temperatures can easily shoot over 1000°C resulting in the explosion of the tank and a calamity. The University of California has estimated that if a storage tank containing 3 million litres of highly radioactive liquid were to explode, an area twice that of Switzerland would be rendered uninhabitable for several decades.

Radiolytic water disintegration produces H_2 and O_2 at a fast rate. If proper ventilation is not provided, the H_2 produced, in the absence of dilution, will reach the lower explosion limit of 4 per cent in a few hours, resulting in an explosion by combining with the O_2.

Packaging of spent fuel: If spent fuel were the primary form of waste, the anticipated packaged waste through the year 2000 would be 2.2×10^6 cubic feet (68,000 cubic metres). If this were stacked as a solid cube, each side would measure nearly 40 metres. About 38,000 megacuries and 175 MW of heat would be produced by this mass of spent fuel.

According to the Swedish project, 1977 Kaernbraenslesaekerhet (KBS), spent fuel should be stored in a water pool in a granite cavern 30 metres below the surface.

After 40 years of storage to dissipate heat, bundles of 500 fuel rods would be loaded in copper canisters with lead and copper covers. Each canister, weighing 18 tonnes, will be transferred to the granite cavern 500 metres below the ground in holes 7.7 metres deep and 1.5 metres in diameter, lined with 40 centimetres of isostatically compressed bentonite.

Radioactive wastes reveal radical changes after few hundred years. First, the heat generation rate decreases by an order of magnitude in the period of 10–100 years and by another order of magnitude in 100–1000 years (the decrease of the heat generation rate depends on the half-life period of the particular radioactive waste). Secondly, the toxicity of HLW needed for 1 GW year electricity decreases by about three orders of magnitude in the first 300–400 years due to the decay of short-lived fission products (1 Giga Watt = 10^9 watts; a measure of consumption of electricity). Toxic levels drop to the level of average ores of toxic elements. After this time, toxicity diminishes slowly, a million years being required for another two orders of magnitude. Thus, the first 300–400 years represent the most critical phase of disposal.

Use of salt mines: The idea originated in West Germany, as salt mines have very little connections with groundwater, thus conferring a high degree of storage security. Asse II, Wolgen Buttel, Germany, stores many small and large caverns filled and sealed with mildly radioactive wastes. By AD 2000, Asse II is expected to store upto 2,50,000 cubic metres of mildly radioactive wastes.

Recent reports of groundwater contamination questions the vulnerability of the system.

Besides salt; granite, basalt and shale have been extensively studied. As a repository should contain and isolate these wastes, site selection involves the consideration of the properties of *host rock*, the hydrologic properties of the site, its tectonic stability, its resource potential and the capability of the site geohydrology, to provide natural barriers to the movement of the waste.

Turning to the sea: UK currently deposits 80,000–90,000 Ci per year of LLRW into the ocean each year, which constitutes 90 per cent of the total waste deposited by Europe (Joyce 1983). Although US abandoned this method in the 1960s, it had deposited about 1,00,000 Ci by then.

Sub-seabed geologic disposal: The abyssal hill regions in the centres of sub-ocean tectonic plates underlying large ocean surface currents, are vastly remote from human settlements, biologically unproductive, have weak and variable bottom currents, and are covered with red clays to a depth of 50–100 metres. The clay has a high cation retention capacity, low permeability, vertical and lateral uniformity; and it becomes increasingly rigid and impermeable with depth (Hollister et al. 1981).

Only about 0.006 per cent of the area of central North Pacific would enable the disposal of HLW by the US till 2040.

CONCLUSION

Hazardous wastes should be disposed of as early as possible and with as little damage to the environment as possible. Currently cost-efficient technology for handling a large number of hazardous wastes is lacking.

REFERENCES

Bond, D.H. (1984) 'At-Sea Incineration of Hazardous Wastes'. *Environmental Science and Technology* 18: 148A–152A.

Forsberg, C.W. (1984) 'Disposal of Hazardous Elemental Wastes'. *Environmental Science and Technology* 18.

Glasstones, S. and W.H. Jordan (1980) 'Nuclear Power and its Environmental Effects'. In *Report of the American Nuclear Society*. Larange Park, Illinois.

Geping, Q. (1982) 'Environmental Protection in China'. *Environmental Conservation* 9.

Gerstly, R.W. and D.W. Albrinck (1982) 'Atmospheric Emissions of Metals from Sewage Sludge Incineration'. *Journal of the Air Pollution Control Federation* 32.

Josephson, J. (1984) 'Hazardous Waste Research'. *Environmental Science and Technology* 18.

Joyce, C. (1983) 'Britain Isolated over Sea-dumping of Nuclear Waste'. *New Scientist* 97.

Health, C.W. (1983) 'Field Epidemilogical Studies of Populations Exposed to Waste Dumps'. *Environmental Health Perspectives* 48: 3–7.

Hileman, B. (1983) 'Hazardous Waste Control'. *Environmental Science and Technology* 17: 281A–285A.

Hinrichsen, D. (1983) 'Europe's Plague of Poisons'. *International Wildlife* 13: 33–5.

Holden, C. (1980) 'Love Canal Residents under Stress'. *Science* 208: 1242–44.

Hollister, C.D., D.R. Anderson and G.R. Heath (1981) 'Sub-seabed Disposal of Nuclear Wastes'. *Science* 231: 1321–26.

Meyer, C.R. (1983) 'Liver Dysfunction in Residents Exposed to Leachate from a Toxic Waste Dump'. *Environmental Health Perspectives* 48: 9–13.

Miller, S. (1981) 'Hazardous Waste Management'. *Environmental Science and Technology* 15: 1413–16.

Moghissi, A.A. (1984) 'Definition of Radioactive Waste'. *Nuclear and Chemical Waste Management* 5.

Pearce, F. and C. Caufield (1981) 'Toxic Wastes: The Political Connections'. *New Scientist* 90: 408–10.

Shen, T.T. (1981) 'Control Techniques for Gas Emissions from Hazardous Waste Land Fills'. *Journal of the Air Pollution Control Association* 31.

Sobrino, F. (1984) 'Hazardous Waste Generators Eye Advances in Incineration'. *Chemical Marketing Reporter* 226.

Thomas, L.M. (1984) 'EPA Fights Hazardous Wastes'. *EPA Journal* 10 (8): 4–7.

White, I.L. and J.P. Spath (1984) 'Low Level Radioactive Waste Disposal'. *Environment* 26.

5

Hazardous Wastes and Their Management—II
Handling, Storage and Transportation

A ny programme for hazardous waste management should not only take care of treatment and disposal, but also of the aspects of safe handling, storage, transportation and risk assessment. It would be appropriate to briefly examine each of these aspects as prerequisites for an integrated hazardous waste management plan.

TRANSPORTATION OF HAZARDOUS WASTES

The discharge of hazardous waste is defined as 'the accidental or intentional spilling, leaking, pumping, pouring, emitting, emptying or dumping of hazardous wastes into air or land or water'. The safe transportation of hazardous materials is an onerous task. The accidental release of hazardous materials may give cause a catastrophe.

Containers used for hazardous waste materials have to conform to the standard safety norms to ensure that inadvertent spillage during transport should not result in any significant release of the materials in to the environment. Further, the packaging should be adequate and effective (Wentz 1989).

Hazardous products are transported in bulk by vessels, tank cars, etc. in containers such as cylinders, drums, barrels, cans, boxes, bottles and casks. The specification of the packaging depends on the nature of the hazardous materials and the strength of the containers.

Bulk transport

Highway transport: Cargo tanks are the main carriers of bulk hazardous material over roads. These are usually made of steel and aluminium alloys. Titanium, Nickel or stainless steel may also be used to construct these tanks. Cargo tanks usually have a lifetime of 8–10 years. Those carrying corrosives have much shorter life spans. Between loads, cleaning has to be done. This further reduces the life span.

Rail transport: The two major classifications of rail tank cars are *pressure* and *non-pressure* for transporting both gases and liquids. These two differ in the type of discharge values, pressure relief systems, and type of thermal shielding.

The commodities most commonly transported by rail are flammable liquids and corrosive materials.

The most common material for construction is steel. Aluminium is the second most widely used metal (Wentz 1989).

Precaution: There is a possibility of a puncture in the pressurised cars carrying flammable material. This may result in excessive heating and consequent expansion of the contents, resulting in an explosion. Safety relief devices have to be provided to avoid this. In addition, top and bottom shelf couplers, which are less risky than relief devices, may be employed. For flammable gases, ethylene oxide and ammonia gas tank cars, installations of head shields is mandatory to provide further protection against coupler damage.

Water transport: The largest bulk containers for water transport are ships, tankers and tank barges. The most common materials that are transported are petroleum products and crude oil. Chemicals such as sulphuric acid, sodium hydroxide, alcohols, benzene and toluene constitute the rest.

The amount of material transported in this manner is large and vessels travel slowly. Sufficient safety measures are therefore taken and statistically this mode is the safest.

All shipments are subjected to strict regulation. The tanker captains and operators should demonstrate a thorough knowledge of pollution laws and regulations, procedures for discharge containment and cleanup, and methods for disposal of sludge and waste materials. Regular inspection of the tankers is essential and this may be the reason for the safety record of water transportaion.

Non-bulk transport

Materials that are transported via non-bulk transport are fibreboard, plastic, wood, glass, fibreglass and metal. Packages within packages are often used for the transportation of hazardous waste materials. For compressed gases, independent units such as steel drums and cylinders are used.

STORAGE

The storage of hazardous wastes is a short term proposal for later collection of material and reprocessing.

Dumping for future recovery is employed in the short term for liquid wastes (Bridgewater and Mumford 1979), such as solvents, which are accumulated until recovery is practical and viable, but long term storage has only been seriously proposed for solid wastes except for radioactive liquid wastes which is a special case (Gaskarth and Bridgewater 1974).

The safety measures to be taken during storage include segregation of incompatible materials (e.g., nitrates and chlorates from carbonaceous materials, and oxidizing acids from organic materials), compliance with limits set for stocks of potentially hazardous chemicals, and proper storage, segregation, handling and supporting of gas cylinders.

Underground storage tanks constitute one of the important methods of storage. Much information is available on the parameters on which the tank owner should focus his resources to minimize overall risk. For example, an older unprotected tank located in the recharge zone of a major public drinking water supply will warrant expedient replacement with a double-walled system. On the other hand, a new tank in a less environmentally sensitive area may be economically retrofitted at a later time to meet minimum requirements. The important parameters to be considered while handling storage tanks are release potential characteristics, tank age, corrosion protection controls, leak detection controls, tank design and soil corrosivity.

The other face of the coin is site-vulnerable data such as soil hydraulic conductivity, surrounding population, aquifer use and proximity to surface waters.

Reconditioned drums as hazardous waste containers

The normal practice of waste disposal is through accumulating it in drums and containers. It is therefore essential that these containers be maintained in good condition to prevent corrosion that will pose a threat to human life and the environment. Frequent inspection is therefore an absolute necessity (Wentz 1989).

The drum reconditioning industry uses the hazardous waste generating unit as a potential source of drums. But there is an inherent long term risk in the supply of used drums for reconditioning.

Regulations

Regulations are suggested to obtain and monitor information from shippers and carriers regarding the nature, mode of transportation and disposal of hazardous waste materials. Licensing, registration and permit requirements enable the state and local governments to achieve this. These in turn are used to target enforcement activities, plan and develop emergency response programmes and optimize routing. The hazardous materials guidelines include procedures for analysing risks associated with the use of alternative routes within a jurisdiction. The risk assessment is based on the chance of an accident happening and the effect that might be felt in the affected zone (Wentz 1989).

Conclusion

Compliance with safety norms and guidelines during handling, storage and transportation of hazardous wastes reduces the risk of disasters to a large extent. Strict adherence to these practices coupled with a good knowledge of risk assessment will go a long way towards making the world a better place to live in.

REFERENCES

Bridgewater, A.V. and J. Gaskarth (1974) *Conservation of Materials.* Harwell: United Kingdom Atomic Energy Agency.

—— and C.J. Mumford (1979) *Waste Recycling and Pollution Control Handbook,* Von Nostrand Reinhold Environmental Engineering Series. New York: Von Nostrand Reinhold Publishers.

Martinez, D.J. (1987) 'Hazardous Waste Disposal'. In *McGraw-Hill Encyclopedia of Science and Technology* 8: 325. New York: McGraw-Hill Inc.

Wentz, C.A. (1989) *Hazardous Waste Management,* Civil Engineering Series. New York: McGraw-Hill International Edition.

6

Disaster Management—I
An Overview

NEED FOR DISASTER FORECASTING AND RISK ASSESSMENT

Chemical process industries often involve reactors, conduits and storage vessels in which hazardous substances are handled at high temperatures and/or pressures. Accidents in such units, caused either by material failure (such as a crack in a storage vessel), operational mistakes (such as raising the pressure, temperature or flow-rate beyond critical limits), or external perturbation (such as damage caused by a projectile), can have serious, often catastrophic, consequences. The most gruesome example of such an accident is the Bhopal gas tragedy of 1984 which killed or maimed over 20,000 people but there have been numerous other accidents in which the death toll would have been as high as in Bhopal if the areas where the accidents took place were not less densely populated (Kletz 1986).

As the density of the industries as well as that of human populations continues to grow everywhere, the risk posed by probable accidents also continues to rise. This is particularly so in the Third World where population densities are very high and more often than not industrial areas are surrounded by dense clusters of neighbourhoods. Further, it is common to have 'industrial areas' or 'industrial complexes' where groups of industries are situated in close proximity to one another. The growth in the number of such industrial areas and in the number of industries contained in each of the areas gives rise to increasing probabilities of 'chain of accidents' or *cascading/domino* effects wherein an accident in one industry may cause another accident in a neighbouring industry which in turn may trigger off another accident and so on.

The science of risk assessment (Khan and Abbasi 1995, Lees 1980a), which has emerged in recent years with ever increasing importance being attached to it, deals with the following key aspects of accidents in chemical process industries:

a) Development of techniques and tools to *forecast accidents*. This is aimed at creating opportunities to rectify problems (of man and materials) *before* any harm;

b) Development of techniques and tools to *analyse consequences* of likely accidents. Such consequence analysis fulfills two objectives:
   – it helps in siting of industries and management of sites so as to minimize the damage if an accident does occur,
   – it provides feedback for other exercises in accident forecasting and disaster management;

c) Development of managerial strategies for 'emergency preparedness' and 'damage minimization' (Khan and Abbasi 1995).

An analysis of major chemical disasters in this century reveals that the following causes and causative factors have been of prime importance:
- runaway reaction, explosion, or fire in the plant
- leaks from pipes, vessels or valves
- design fault leading to accidental mixing of chemicals
- improper maintenance
- storage of unacceptably large quantities of hazardous and vulnerable chemicals
- accidents during transportation
- inventory control failures
- improper toxic waste disposal-practices
- lack of security or preparedness for disaster management (Kletz 1986, Fawcett and Wood 1982, Segraves 1991) .

In this chapter, case studies of typical process industry

accidents are described. The essence of the science of risk assessment (also called risk analysis), which developed as a result of these experiences, is presented to prevent or minimize the impact of the accidents.

CASE HISTORIES OF MAJOR CHEMICAL DISASTERS WITH SPECIAL REFERENCE TO THEIR IMPACT ON ENVIRONMENT

The locations, responsible chemicals, type of disaster and casualties due to some major process industry accidents in the last three decades are summarized in table 6.1.

We recount below, in brief, some of the major chemical disasters that typify what can go wrong and how; especially the 'chain reactions' or the *cascade/domino* effects where one explosion leads to another which in turn triggers yet another, rapidly leading to uncontrollable situations.

Disasters at the industry site

The Flixborough disaster: The Flixborough Plant of Nypro Limited, UK, was built for the production of caprolactum which is the basic raw material for the production of Nylon 6 (Lees 1980a and 1980b, Kletz 1990). Cyclohexanol, an essential raw material was produced by the oxidation of cyclohexane. The last named chemical, which in many of its properties is comparable with petrol, had to be stored. More importantly, large quantities of cyclohexane had to be circulated through the reactors under a working pressure of about 8.8 kg/cm^2 and a temperature of 155°C. The reaction is exothermic; any escape of cyclohexane from the plant would therefore be dangerous. The cyclohexane plant at Flixborough consisted of a stream of six reactors in series (Figure 6.1) in which cyclohexane was oxidized to cyclohexanone and cyclohexanol by air-injection in the presence of a catalyst.

On the evening of 27 March 1974, it was discovered that cyclohexane was leaking from reactor number 5. The following morning an inspection revealed that the leak had extended by about 6 feet. This was a serious state of affairs and a meeting was called where a decision was taken to remove reactor 5 and to install a bypass assembly to connect reactor 4 directly to reactor 6 so that the plant operation could continue.

Table 6.1 Some major accidents in the process industries (1970–96)

Year	Location	Chemical	Event	Deaths
1971	Houston, USA	Vinyl chloride monomer	BLEVE	1
1972	Brazil	Butane	UVCE*	37
1972	West Virginia, USA	Gas	CVCE	21
1973	Potchefstroom, South Africa	NH_3	Toxic release	18
1973	State Island, USA	LNG	Fire in empty storage tank	40
1974	Flixborough, UK	Cyclohexane	UVCE	28
1976	Seveso, Italy	Tetra chlora dibenzo paradioxin	Toxic release	–
1977	Columbia	NH_3	Toxic release	30
1978	Waverly, USA	Propane	BLEVE	12
1981	Montanas, Mexico	Chlorine	BLEVE	29
1982	Spencer, USA	H_2O	BLEVE	7
1983	Houston, USA	Methyl bromide	BLEVE	2
1984	Mexico city, Mexico	Propane	BLEVE	650
1984	Bhopal, India	Methyl-iso-cyanate	Toxic release	2500
1986	Kennedy Space Center, USA	Hydrogen	BLEVE	7
1987	Piper Alpha, UK	Hydrogen	Explosion & fire	167
1988	Illinois, USA	Propane	BLEVE	15
1989	Pasadena, USA	Iso-butane	BLEVE	23
1989	Antwerp, Belgium	Ethylene oxide	UVCE	20
1989	USSR	Ammonia	Explosion & toxic release	7
1989	Pasadena, USA	Ethylene	Explosion	23
1990	Thane, India	Hydrocarbon	Fire & explosion	35
1990	Porto de Leixoes, Portugal	Propane	Fire & explosion	14
1993	Panipat, India	Ammonia	Explosion & toxic release	3
1995	Gujarat, India	Natural gas	Fire	@
1996	Mumbai, India	Hydrocarbon	Fire	@

BLEVE— Boiling liquid expanding vapour explosion
UVCE — Unconfined vapour cloud explosion
CVCE — Confined vapour cloud explosion
@ — Factory badly damaged; death total not known

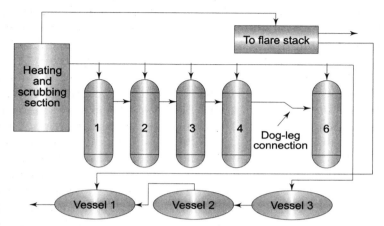

Fig. 6.1 Simplified flow diagram of the sequence of reactors used in the cyclohexane oxidation unit at Flixborough. Note the dog-leg connection put as an ad-hoc measure between the fourth and the sixth reactors, which led to the disaster

The openings to be connected on these reactors had a diameter of 28 inches, but the largest pipe which was available on site and which was suitable for the by pass had a diameter of 20 inches. The two flanges were at different heights so that the connection had to take the form of a dog-leg of three lengths (Figure 6.2). Calculations were done to check that the pipe had a large enough cross-sectional area for the required flow and that it was capable of withstanding the pressure as a straight pipe.

But no calculations took into account the forces arising from the dog-leg shape of the pipe; no drawing of the by-pass pipe was made other than in chalk on the workshop floor; and no pressure testing was carried out either on the pipe or on the complete assembly before it was fitted. A pressure test was performed on the plant, after installation of the bypass, but the equipment was tested to a pressure of 9 kgF/cm (Khan and Abbasi 1995). Further, the test was pneumatic, not hydraulic (Lees 1980a).

The plant was restarted. Initially the bypass assembly gave

no trouble. On 29 May 1974 the bottom valve on one of the vessels was found to be leaking. The plant was again shut down for repairs, and restarted on 1 June. A sudden rise in pressure upto 8.5 kgF/cm² occurred early in the morning when the temperature in reactor 1 was only 110°C and lower in other reactors. Later that morning, the pressure reached was 9.1–9.2 kgF/cm².

During the late afternoon an event occurred which resulted in the escape of large quantities of cyclohexane. This event was the rupture of the dog-leg shaped bypass system. It was perhaps aided by a fire on a nearby 8-inch pipe. The fugitive cyclohexane soon caught a spark, and there was a massive unconfined vapour-cloud explosion. The blast and the fire destroyed the cyclohexane plant as well as several other plants in its vicinity.

Of those working on the site at the time, 28 were killed and 36 others suffered injuries. Outside the plant, injuries and damage were widespread but no one was killed. Of the 28 people who died, 18 were in the control room. Some of the bodies suffered severe damage from flying glass. The main office of

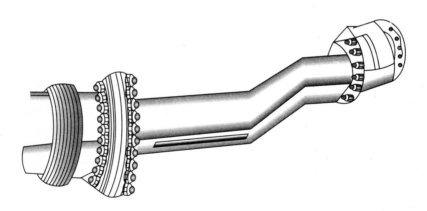

Fig. 6.2 Detail of the 20" bypass pipe and scaffolding which precipitated the Flixborough disaster (see Fig. 6.1)

the factory was demolished by the blast of the explosion. Mercifully the accident occurred on a Saturday afternoon when the offices were not occupied.

Property damage extended over a wide area, and a preliminary survey showed that 184 houses and 167 shops and factories had suffered to a greater or a lesser degree.

Seveso disaster: On the morning of Saturday, 10 July 1976, a safety-valve vented on a reactor at the Icmesa Chemical Company at Seveso, a town of about 17,000 inhabitants some 15 miles from Milan (Italy). A white cloud drifted over part of the town and heavy rainfall brought the cloud to earth (Fawcett and Wood 1982, Lees 1980b). The emission occurred from a reactor producing trichlorophenol, which is used to make a bactericide, hexachlorophenol, and a herbicide, 2,4,5-trichloro phenoxy acetic acid. The reactor also contained the chemical generally referred to as TCDD (2,3,7,8-tetrachloro dibenzo paradioxin). This substance was not an intended reaction product but an undesired byproduct. An estimated 2 kg of TCDD was released although this estimate is necessarily approximate.

In normal operations the amount of TCDD made in the reactor was small, but on this occasion the reactor had got out of control. The contents had got overheated and the safety valve had vented. The high temperature in the reactor favoured the production of abnormal quantities of TCDD.

In the immediate area of the release the vegetation was contaminated and animals began to die. On the fourth day a child fell ill and on the fifth day civil authorities declared a state of emergency in Seveso. An area of about 2 square miles was declared as contaminated and people were asked to avoid contact with the vegetation and edible products from this area. The contaminated area was later sought to be closed completely. On 27 July the first evacuation of about 250 people took place. By the end of July, 250 cases of skin infection had been diagnosed. About 100 people were told to evacuate their homes and about 2000 people had to be given blood tests. In early August, it was found that the area contaminated was about 5 times larger than was originally thought (Lees 1980b).

There have been accidents involving TCDD release prior to the Seveso disaster. At Ludwigshafen Germany, fifty-five people were exposed when there was accidental TCDD release in 1953, and many developed severe symptoms of TCDD poisoning. Various measures were taken to decontaminate the plant building, including the use of detergents, the burning off of the surfaces, the removal of insulating material and so on, but these were not effective and eventually the whole building had to be destroyed. In another accident at the Philips-Duphar plant near Amsterdam in 1963, a leak of 0.03–0.2 kg of TCDD occurred. About 50 people were involved in cleaning up the leakage, of whom four subsequently died, and about a dozen suffered occasional skin troubles. The plant was sealed for ten years and then dismantled from the inside brick by brick, the rubble was embedded in concrete, and the concrete blocks were sunk in the Atlantic. Five years later yet another accident involving TCDD release occurred at Bolsover. It involved a runaway reaction in a trichlorophenol reactor, similar to one that later occurred at Seveso. The reaction temperature reached 250°C, the reactor exploded and the supervising chemist was killed. The plant was closed down, and then reopened after two weeks when it appeared that the workers exposed had suffered no ill effects. But within seven months, 79 people complained of TCDD symptoms. The plant was dismantled and buried in a deep hole. But the story did not end there; three years later contractors on the site developed TCDD symptoms. The only apparent possible source of contamination was a metal vessel which had been thoroughly cleaned and subjected to sensitive testing.

The lesson that emerged from Seveso was that pressure-relief valves on plants handling highly toxic substances should not discharge to the atmosphere but to a closed system. If this lesson had been learnt by Union Carbide, the worst disaster in the history of chemical process industries, the Bhopal tragedy, could have been averted.

Bhopal disaster: The worst ever disaster in the history of the chemical industry occurred in Bhopal, India, on 3 December 1984. A leak of methyl isocyanate (MIC) from a chemical plant, where it was used as an intermediate in the manufacture of a pesticide, spread beyond the plant boundary and caused death by poisoning of over 2,500 people while injuring about 10 times as many (Kletz 1986, Shrivastava 1987).

Methyl isocyanate boils at about 40°C at atmospheric pressure. According to press reports, the contents of the storage tank overheated and boiled, causing the relief valves to lift. The discharge of vapour, about 25 tonnes, was too great for the capacity of the scrubbing system. The escaping vapour spread beyond the plant boundary where a shanty town had sprung up. The cause of the overheating was contamination of the methyl isocyanate, by water or other materials, and several possible mechanisms were suggested. According to some reports, cyanide was produced. If contamination did occur, it reinforces the message that hazard and operability studies should be carried out on all new plant designs. They provide an effective means of showing up ways in which contamination can occur. Further more if a risk analysis (specifically maximum credible accident analysis) had been conducted, it would have indicated that in the event of a MIC leak the scrubbing system would be inadequate. This would have enabled Union Carbide to install better emergency handling systems.

PEPCON explosion: On 4 May 1988, a massive explosion destroyed a plant of Pacific Engineering and Production Company, (PEPCON), near Henderson, about 12 miles south of Las Vegas, USA (Fawcett and Wood 1982).

PEPCON was one of only two plants in the USA that produced ammonium perchlorate (AP); the other was the Kerr-McGee plant, also located in Henderson about 2 miles from the PEPCON plant. PEPCON reportedly produced about one-third of the AP used as an oxidizer and propellant in solid, composite rocket fuels for NASA's space shuttle and missiles.

Although a fire started the PEPCON explosion, the cause of the fire was not easy to explain. After the explosion, PEPCON blamed the fire on a leaking underground pipeline of the Southwest Gas Company that traversed PEPCON'S property.

But the natural gas pipeline had been installed about 10 years before PEPCON's plant, and although ruptured, it only contributed to the fire and heat required to detonate the second and the largest explosion.

The fire was also attributed to a welder's torch but one of the reports absolved the welder of the blame. Some blamed the batch dryer's fibre-glass insulation which had a history of AP spills into the combustible insulation.

The following were the tell-tale conditions in and around PEPCON:

- lack of proper storage
- combustible fibre-glass insulation and sources of fire
- glass panel walls in the batch house
- inadequate spacing between adjacent process vessels and product storage tanks
- no alarm to warn plant personnel, fire departments or Henderson's other citizens
- no dependable fire-fighting arrangement with sprinklers and deluge system
- no modern, dependable, radio system to back-up damaged telephone lines needed to call for help, coordinate response teams and warn the community
- lack of an effective emergency response plan at PEPCON, within the surrounding industrial complex and within the town of Henderson.

The explosion caused about $ 100 million in damages to the surrounding community and completely destroyed a neighbouring marshmallow plant. About 350 people were injured. Two people died—the plant manager and the controller.

Piper Alpha explosions: It has been said that those who forget history are condemned to repeat it. Piper Alpha was given a painful reminder of this adage in 1988 when an explosion killed 167 people and caused massive monetary losses (Kletz 1991a).

The firm had received several warning signals before but had not given them the attention they deserved. In 1928, when one of the factory's low-pressure gas main was being modified, it had been isolated from a gasholder containing hydrogen by a closed valve. This valve, unknown to those concerned, was leaking. Soon the leaking gas was accidentally ignited. There was a loud explosion and flames appeared at a number of joints on the main. One man was killed. The source of ignition was a match struck by one of the fitters so that he could see to take some measurements. He thought it was safe to strike a match as he had been told that the area was gas-free. Even if the fitter had not lit his match, some other source of ignition might have turned up as always seems to happen when an explosive mixture is let loose. The leak, rather than the match, was therefore the cause of the explosion.

In 1967, on the same site, a large pump was dismantled for repair. When a fitter removed a cover, hot oil came out and caught fire as the suction valve had been left open. The oil was at about 280°C above its auto-ignition temperature; ignition was quick and inevitable. Three men were killed and the plant was destroyed.

The third chapter in this saga occurred in 1987 near the site of the first incident. A hydrogen line had to be repaired by welding. The hydrogen supply was isolated by closing three valves in parallel (one of which was duplicated). The line was then purged with nitrogen and, before the welding started, it was tested at a drain point to confirm that no hydrogen was present. When the welder struck his arc an explosion occurred and he was injured. The investigation showed that two of the isolation valves were leaking. It also showed why the hydrogen was not detected at the drain point: it was at a low level and air was drawn through it into the plant to replace the gas leaving through a vent. The source, of ignition was sparks from the welding unit because the unit's return lead was not securely connected to the plant.

The accident report recommended that slip-plates should be used in future for the isolation of hazardous materials. There was no reference to the earlier incidents which were probably not known to the author.

As no adequate lessons seemed to have been learnt from the repeated accidents, the worst was still in store. It occurred in 1988 which, as mentioned above, took a far heavier toll of life and property than all the previous explosions put together.

According to the official report, a pump relief valve was removed for overhauling and the open end of the pump had blanked. Another shift engineer, not knowing that the relief valve was missing, started up the pump. The blank was probably not tight and the condensate (light oil) leaked past it and exploded.

The post-disaster enquiry report concluded that on a balance of probabilities the leakage of condensate was from a blind-flange assembly which was not leak-tight (Kletz 1991b).

One notices that in the Piper Alpha series of tragedies, the operating staff always displayed uncertain commitment to working as per the written procedures; and that often the procedures were knowingly and flagrantly disregarded.

The Phillips explosion: The explosion at the Phillips high density polyethylene plant in Pasadena, Texas, on 23 October 1989 is one of the worst industrial accidents of the last 10 years.

The immediate cause was simple: a length of pipe was opened up to clear a choke without bothering to see that the isolation valve (which was operated by compressed air) had not been duly closed. The air hoses which supplied power to the valve were connected up the wrong way round so the valve was open when its actuator was in the closed position. Identical couplings were used for the two connections so it was easy to reverse them. According to the company procedure they should have been disconnected during maintenance but they were not. The valve could be locked open or closed but this hardly mattered as the lock was missing. The explosion occurred less than 2 minutes after the leak started and two iso-butane tanks exploded 15 minutes later. The explosive force was equivalent to 2.4 tonnes of TNT; 23 employees were killed and over 130 injured. Some 40 tonnes of ethylene gas leaked and exploded (Kletz 1991b).

Disasters during transportation

Explosion of High Flyer and Grand Camp ships, Texas City, USA: At about 8.10 am on 16 April 1947, fire was observed around some of the ammonium nitrate fertilizer bags loaded on board the ship *Grand Camp* in the harbour at Texas City, USA. There were 880 tonnes of ammonium nitrate in that hold and a further 1400 tonnes in another hold. Frantic efforts were made to extinguish the fire, but the quantity of water used initially was too small and by the time a hose line had been connected to supply larger quantities of water, the fire was so well-established that the crew was ordered to abandon the ship. At 9.15 am there was a loud explosion resembling a thunderclap and *Grand Camp* disintegrated killing all persons in the dock, including firemen and spectators.

Another ship *High Flyer* which also had ammonium nitrate on board, was 700 feet away and was blown free of its anchors. On account of the danger of another explosion, volunteers could not be found to move *High Flyer* out of the burning area.

At 6 in the evening, *High Flyer's* sulphur cargo ignited. At 1.10 am the next day, the ship was ripped apart by the expected explosion.

Insurance inspection revealed blast damage in excess of 3,300 dwellings, 130 business buildings, more than 600 automobiles, and some 360 boxcars. Some 125 storage tanks, ranging from 25,000 to 80,000 barrels in capacity, were substantially damaged either by missiles or intense fire.

There were 552 deaths and over 3000 injuries in a community of some 15000 people (Fawcett and Wood 1982, WHO 1990).

Derailment of LPG carrying tank cars, Crescent City, USA: On 21 June 1970 a railway train passing through the streets of Crescent City, Illinois, USA was accidentally derailed. Nine rail tank cars loaded with liquid petroleum gas (LPG) were among the derailed coaches. The force of the derailment propelled the 27th coach over the other derailed coaches. One of its couples struck the tank of the 26th car and punctured it. Propane was released and caught fire.

The safety valves of the other tank coaches opened and released propane, which fed the fire. At about 7.33 am the 27th coach exploded. Four fragments of the coach were hurled in different directions for distances of 300, 600 and 750 feet. At about 9.40 am the 28th coach exploded hurling one fragment which eventually stopped after rolling to a distance of 1600 feet. At about 9.45 am the 30th coach exploded and at about 10.55 am the 32nd and 35th coaches.

The main mechanism of damage in this incident was the rocketing of burning tank cars which hurled massive fragments and fresh fire balls upto 850 feet from the site of the derailment. The fire balls were several feet tall.

There were no fatalities but 66 persons were injured. There was extensive damage to property (Lees 1980a, 1980b).

BASIC COMPONENTS OF A HAZARD-CONTROL SYSTEM

Very broadly, the basic components of a risk-reduction or hazard-control system are:

1. *Incident reduction*
   - operating procedures
   - mechanical integrity of equipments
   - process safety information
   - process hazard analysis
   - training
   - safe work practices, and

2. *Incident management*
   - emergency plan
   - incident investigation (Khan and Abbasi 1995, WHO 1990, World Bank 1995).

These objectives are accomplished in the following ways:

a) Identification of major hazardous installations: It is necessary to identify the installation which, according to the definition, may fall within the criteria set for the classification of major hazardous installations (Kandal and Avni 1988, Kumar and Rao 1991).

b) Information pertaining to the installations: Once the major hazardous installations have been identified, additional information should be collected on their design and operation. In addition, such information must also describe all other hazards specific to each installation (Kandal and Avni 1988, Kumar and Rao 1991).

c) Actions by government authorities: Licence to an existing industry to continue or expand, or to a proposed industry to start, should be given only after a careful risk assessment study. Thereafter, periodic inspection and strict enforcement of legislation by the governments should be done to preempt disasters. Land-use planning can also appreciably reduce the potential for disaster and should come under close governmental supervision. Insistence for the training of factory inspectors in industrial safety is also an important role of the government (World Bank 1985, Lesins and Joseph 1991).

d) Actions inside the zones of industrial activity: The primary responsibility of management of a plant is to operate and maintain a plant in a safe manner. Technical inspection, maintenance, plant modification, training and selection of suitable personnel must be carried out according to sound procedures in addition to the preparation of the safety report. Each accident, however minor or apparently trivial, must be thoroughly investigated to identify its causes. Often an apparently trivial accident may be symptomic of a major lacuna and can provide the warning signal which, if needed, can help in averting great tragedies in future (Kumar and Rao 1991, Uitenham and Munjal 1991).

e) Emergency Planning (EP): EP aims at the reduction of the consequences of major accidents, and assumes that absolute safety cannot be guaranteed. In setting up an emergency off-site programme, a well-structured and clear plan is one which is based on a well-prepared safety report and which can be quickly and effectively employed when a major accident occurs (Uitenham and Munjal 1991, Hawks and Merien 1991).

Techniques of risk assessment

In striving for improvement of the safety of chemical process systems numerous formalized procedures for plant safety analysis have been developed (Kletz 1986, WHO 1990, Kandal and Avni 1988). They are:

- PHA   Preliminary hazard analysis
- HAZOP   Hazard and operability studies
- HAZAN   Hazard analysis
                  a) Fault-tree analysis
                  b) Fatality analysis
- MCAA   Maximum credible accident analysis.

As HAZOP and HAZAN are often confused by those unfamiliar with them, it may be of use to spell out the differences between them: Hazard is *identified* in HAZOP studies whereas it is *assessed* in HAZAN, HAZOP is essentially *qualitative* while HAZAN is *quantitative*, and HAZOP studies lead to HAZAN; in other words HAZOP is a precursor to HAZAN.

Preliminary hazard analysis (PHA)

PHA is performed as the first step in risk assessment. It starts by identifying *likely* accidents which may involve toxic, flammable and explosive materials. The procedure specifies system elements or events that *can* lead to hazardous conditions. Once a hazardous system has been identified, the event that may lead to the accident is specified. Events such as 'the formation of an explosive atmosphere outside or inside a storage vessel' or 'the release of a toxic gas' will need to be examined, so as to identify the components of the plant that *can* cause the accident. The components which include storage tanks, reaction vessels, pipes, pumps, stirrers, relief valves or other systems, are then singled out for a more detailed examination by other evaluation methods such as HAZOP (WHO 1990).

Since the PHA is fast and cost effective, and identifies key problems, hazard evaluation should always start with this method. Its results indicate which system or procedure requires further analysis and which is of less interest from the point of view of major hazards. In this way it is possible to limit the evaluation to key problems, thus avoiding unnecessary effort during the later steps (Kletz 1986, Shrivastava 1987).

HAZOP (hazard and operability) and 'What if' studies

The results of a PHA may show that a more detailed study is required of a particular section of the plant which presents a significant hazard to operators or to the environment. For this purpose, a HAZOP study is required to identify systematically the possible ways in which the system could fail.

The objectives of HAZOP studies are:

- to identify those areas of design which may possess a significant hazard potential
- to identify and study features of the design which influence the probability of the occurrence of hazardous incidents
- to familiarize the study team with the design information available
- to ensure that a systematic study is made of the areas of significant hazard potential
- to identify pertinent design information not currently available to the team
- to provide a mechanism for feedback to the client, of the study team's detailed comments.

A flow-chart of a HAZOP procedure is presented in figure 6.3.

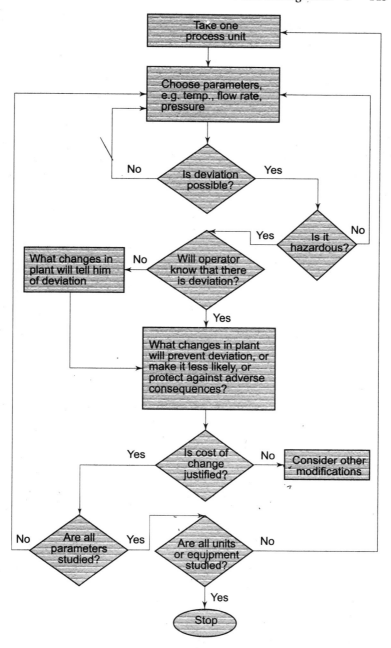

Fig. 6.3 A typical component of a HAZOP study

Hazard analysis (HAZAN)

The method of hazard assessment called 'hazard analysis', effectively recognizes four situations with respect to the probability of an event constituting a risk: (i) the probability of the event is low; (ii) the probability of the event can be *made* low by the application of a standard or a code of practice; (iii) the probability of the event can be made low by the application of measures which can be shown by a simple risk-analysis to be of equivalent safety to the normal standard or code of practice; (iv) the probability of the event is assessed quantitatively and is reduced to conform to the risk criteria by measures indicated by the analysis. For simple cases, use is made of a simple *fatality analysis* (FA) while for complex cases, it is necessary to resort to a detailed *fault-tree analysis* (FTA) (Kandal and Avni 1988). Both options are described below.

Fatality analysis: Attempts made to decide a risk criterion tend to follow two main approaches. One is to determine the risk to which it is permissible to expose an individual. The other is to determine the maximum sum of money which may have to be spent in order to avoid fatalities (Kletz 1986, Lees 1980a).

A simple risk analysis is carried out to demonstrate equipment safety. The spirit in which the analysis is done is similar to that of the early work on operational research. The situation is described and studied by a simple model. The implications of the model are then explored, initially by simply using estimation and then, if the need is indicated, field data. The decision is then made. Most treatments of acceptable risk primarily deal with the risk of death. Data on fatalities are recorded. Fatal Accident Frequency Rate (FAFR) is defined as the number of deaths per 10^8 exposed hours; this corresponds roughly to the number of deaths over a working-life time among a group of thousand workers. This is also referred to as Fatal Accident Rate (FAR). Another index which is useful in relation to general fatality risks is the death rate per annum (Kletz 1990, World Bank 1985).

The risk 'r' to an individual may be computed as

$$= \frac{1}{N} \sum_{i=1}^{n} x_i F_i$$

where x_i is the number of deaths in a particular type of potential accident,

F_i is the frequency of such an accident,

n is number of types of potential accidents, and

N is total number of people at risk (Khan and Abbasi 1985, Lees 1980a).

The calculation is concerned with the risk to an individual. The risk to a member of the public may also be assessed using the above equation. In doing this it is necessary to bear in mind the risk from other sources to which the individual may be exposed.

Fault-tree analysis: Fault-tree analysis is a deductive method which is normally used in a quantitative way, although it requires as an initial step a qualitative study of the system under consideration, just as any method of system-analysis does. After defining the undesired event, its logical connections with the basic events (failure of technical components) of the system are searched for and the result of this search is represented graphically by means of a fault-tree (figure 6.4). The tree reflects the outcome of the qualitative part of the analysis, in which questions of the type 'how can it happen' are answered. These serve to identify, firstly, the process functions and subsystems such as cooling or electric supply whose failure causes the undesired event and then connects those failures successively with the basic events. In figure 6.4 the concept of fault-tree analysis is explained using the simple illustration of the process of lighting two bulbs with a battery.

The logical connections in the fault-tree analysis are generally represented by two types of gates, the 'OR' and the 'AND'. In the case of the 'OR' gate, any one of the entries is capable of producing the output event, while the output event of the 'AND' gate occurs only if all its entry events are fulfiled. Sometimes in addition, 'NOT' gates are used which convert the entry into its opposite. Two states of the basic event are normally admitted. They are either true or false (e.g., failed or functioning in the case of technical components), which implies two possible states for the undesired event, its occurrence and its non-occurrence. The two states are adopted with certain probabilities which in the

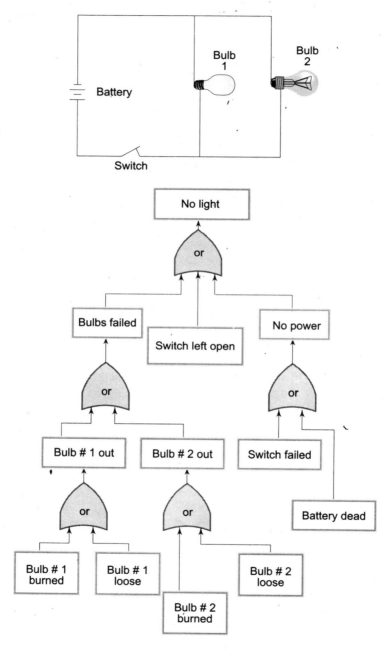

Fig. 6.4 An illustration of the concept of fault tree analysis

case of technical components are generally obtained for each type of component by evaluating the operating behaviour of a large number of similar components. Applying these probabilities to the basic events of a fault-tree, the probability of the undesired event may be calculated.

Generally speaking, the fault-tree approach, which is based on the search for circumstances which make the system fail, may be regarded as an antithesis of the design process which strives to establish the conditions of system functioning. This is why it proves particularly useful in detecting design flaws both at the qualitative and quantitative levels of investigation.

In order to carry out a fault-tree analysis the following steps are required:

- familiarization with the system using process description, piping-instrumentation diagrams, and information obtained from the plant personnel
- definition of the undesired and initiating events using devices such as material information, checklists, and historical evidence
- development of the fault-tree
- obtaining probabilities for the failure of technical components and human error
- evaluation of the fault-tree
- analysis of the results, and articulation of proposals for system improvement, if necessary, for any changes of the fault-trees, and evaluation of the renewed fault-tree (Fawcett and Wood 1982, Kandal and Avni 1988).

Maximum credible accident analysis

Risk assessment is complete only if the consequences of a possible accident are known. For this reason, the last step of a hazard assessment is to analyse the consequences that a potential major accident could have on the plant itself, on the employees, on the neighbourhood and on the environment. The results of the analysis are used to determine which protective measures, such as fire fighting systems, alarm systems, or pressure-relief systems, have to be installed. An accident consequence analysis should contain the following:

- a description of the accident (tank rupture, pipe rupture, failure of safety valve, fire)

- an estimate of the quantity of material released (toxic, flammable, explosive)
- a calculation of the dispersion of the material released (gas or evaporating liquid), and
- an estimate of the effects (toxic, heat radiation, blast wave).

In order to identify the undesirable events, it is essential to construct an accident scenario of possible events. The next step is mathematical modeling of chemical discharge effects based on known physical phenomena. Table 6.2 lists some of the various mathematical models employed for the purpose (Khan and Abbasi 1995).

Table 6.2 Some mathematical models for maximum
credible accident analysis

	Phenomenon	Applicable models
1	Outflow	
	– liquid	Bernoulli flow equation
	– two-phase mixture	Fauske/Cude model
	– gas	gas flow equations
2.	Behaviour immediately after release	
	– spreading liquid release	Spreading liquid model
	– jet dispersion	Simple jet model
	– adiabatic expansion	Two stage expansion model
3.	Dispersion	
	– heavy gas	Cox and Carpenter dense gas dispersion model
	– natural gas	Advanced PPT based model
	– buoyant gas	Birggs Plume rise model
4.	Heat radiation	
	– pool fires	Classical empirical equations employed to determine burning rate, heat radiation and incident heat.
	– jet fires	Jet dispersion model
	– fire balls	Empirical correlation of fireball radius based on American Petroleum Institute model
5.	Explosion	
	– BLEVE (Boiling liquid expanding vapour)	Fireball and radiation models
	– explosion	Over pressure models
	– vapour cloud explosion	Deflagration models

MCAA can be applied to individual industries as well as a cluster of industries. An industrial area where several chemical process industries such as petroleum refinery, fertilizer or synthetic organic chemicals are involved, is considered. These industries pose a risk not only to the resident population outside the industry limits but also to the neighbouring industries by causing 'cascade effects'. MCAA provides an opportunity to assess the potential of various chemical plants in a cluster of these industries to create unsafe situations to the surroundings and the environment. The analysis enables identification of those parts of the facilities for which more detailed safety study will be necessary to assess the risk potential. In an MCAA, the worst accident which is still plausible is considered. The selection of accident scenarios depends on engineering judgment, accident description in the past, and expertise in the field of hazard analysis studies.

The steps followed in a MCA analysis are:

a) a study of the plant and process information obtained for the cluster of industries;

b) preparation of an inventory of major chemical storages and ranking them on the basis of their hazardous properties and storage quantities;

c) identification of potential hazardous plant section and representative failure cases from the processing units;

d) visualization of the chemical release scenarios;

e) effect and damage calculation from the release cases through mathematical modeling;

f) shortlisting of maximum credible accident scenarios.

EMERGENCY CONTROL AND DISASTER PLANNING

An emergency cannot always be prevented but it can be controlled within limits and its effect can be minimized by using the best resources available at the time. The objectives of emergency planning are:

- to provide resources and methods for effective control of emergencies arising out of leakage, explosion, and/or fire due to hazardous chemicals

- to prevent the emergency from turning into a disaster

- to prevent damage to property, people and environment
- to arrange for effective rescue operations and treatment of casualties
- to bring the situation back to normal in the least possible time
- to provide authoritative information to news-media and government agencies, and
- to minimize panic amongst the general public, avoiding exploitation or exaggeration of the situation by any agency.

The management of an industry should first evaluate the basic and distinctive requirements of each site for handling the foreseen emergencies or hazards. The needs to evaluate will be different and depend on:

- size and nature of works within the factory limits
- the location of the factory-site and the plant-site
- the nature of chemicals, and processes
- number of people employed in the factory premises and the availability of resources.

If the emergency is major it is necessary to effect a progressive and total evacuation of all workers including visitors, contractors and other personnel. The management must ensure proper training of the staff and the workers for handling the emergency and they should be able to carry out their specified tasks efficiently at the time of emergency.

Education of the general public is one of the most important and difficult tasks. They should be educated to take precautions at the time of leakage of toxic and inflammable chemicals. Growth of trees and plants around the factory premises would help to some extent in absorbing toxic gases and shock waves (Velan and Abbasi 1995).

TYPICAL SEQUENCE OF STUDIES NORMALLY CONDUCTED FOR RISK ASSESSMENT OF AN INDUSTRY

1. *Introduction—study of the manufacturing process*
2. *Hazard identification*:
   - classification of major hazardous substances
   - identification of major hazardous installations
   - past-accident data analysis

- visualization of all likely accident scenarios
- event-tree analysis to define the outcome of accidents
- short-listing of maximum credible accident scenarios.

3. *Maximum credible accident analysis*
   Use of mathematical models to work out:
   - rates of hazardous releases and intensities of toxic gas releases/explosions/fires
   - mode of dispersal of toxic gases/fireballs/BLEVEs (boiling liquid expanding vapours)/pool fires, etc.
   - vulnerability studies or the potentialities of damages.

4. *Consequence analysis*
   Use of mathematical models to:
   - estimate probabilities of predicted consequences
   - do quantitative risk assessment—assess group or societal individual risk, voluntary or involuntary risk etc.
   - carry out a scenario-wise consequence analysis.

5. *Preparation of disaster management plan*

6. *Preparation of emergency preparedness plan.*

REFERENCES

Fawcett, H.H. and W.S. Wood (1982) *Safety and Accident Prevention in Chemical Operations.* New York: John Wiley.

Hawks, J.L. and J.L.Merien (1991) 'Create a Good PSM System'. *Hydrocarbon Processing* (August), pp. 29–31.

Kandal, A. and E. Avni (1988) *Engineering Risk and Hazard Assessement.* Vol. 1. Florida: CRC Press Inc.

Khan F.I. and S.A. Abbasi (1995) 'Risk Analysis: A Systematic Method of Hazard Assessment and Control', Journal of Industrial Pollution Control 11: 89–97.

Kletz, T.A. (1986) *What Went Wrong: Case Histories of Process Plant Disasters.* London: Gulf Publications.

—— (1990) 'Process Safety—An Engineering Achievement', *Proceeding of Institution of Mechanical Engineers, IMechE: UK,* 25: 11–15 (London).

—— (1991a) 'Piper Alpha: Latest Chapter in a Long Story'. *Journal of Loss Prevention in Process Industries* 4: 271–7.

—— (1991b) 'The Phillips explosion: This is Where I Came In'. *Chemical Engineer* 493: 42–7.

Kumar, A. and H.G. Rao (1991) 'Software for Regulatory Compliance of Chemical Hazards'. *Environmental Progress* 9: 1–7.

Lees, F.P. (1980a) *Loss Prevention in the Process Industries.* Vol. 1. London: Butterworths.

—— (1980b) *Loss Prevention in the Process Industries.* Vol. 2. London: Butterworths.

Lesins, V. and J.M. Joseph (1991) 'Develop Realistic Safety Procedures for Pilot Plants'. *Chemical Engineering Progress* (January), pp. 39–45.

Segraves, R.O. (1991) 'Learn Lessons From PEPCON Explosions'. *Chemical Engineering Progress* 87: 61–5.

Shrivatsava, P. (1987) 'Industrial Crises'. *Chemical Engineering World* (August), pp. 11–19.

Uitenham, L. and R. Munjal (1991) 'Choose the Right Control Scheme for Pilot Plants'. *Chemical Engineering Progress* (January), pp. 31–5.

Velan M. and S.A. Abbasi (1995) 'Modelling of Greenbelts'. Proceedings of the training programme, *Environmental Pollution and Its Control.* Pondicherry: Pondicherry University, May.

WHO (1990) *Major Hazard Control: A Practical Manual.* Geneva: International Labour Office.

World Bank, Office of Environmental Affairs (1985) *Manual of Industrial Hazard Assessment Techniques.* London: Technica Ltd.

7

Disaster
Management—II
A Case Study

*I*n *the previous chapter we have given an overview of the importance of risk assessment and the various methodologies developed for the purpose so far. We now present a case study in which one of the methodologies developed by us has been employed for conducting risk assessment. The case study pertains to the storage units of a typical chemical industry engaged in the manufacture of epichlorohydrin. We have conducted the study using a comprehensive software package MAXCRED (maximum credible accident analysis) recently developed by us.*

INTRODUCTION

As discussed in the preceding chapter, serious accidents can take place in chemical process industries, in the form of explosions, fires and release of toxic chemicals. Such accidents can take a heavy toll in terms of loss of property and human lives. If the accidents involve release of large quantities of toxic chemicals, they can also contaminate large areas and render them useless for several years (Abbasi and Venilla 1994, Khan and Abbasi 1994).

Accidents in process industries are generally caused by the following factors.

a) cracks in storage or reaction vessels leading to rupture
b) malfunctioning of control equipment such as valves

c) other types of failures in instruments and piping
d) human errors
e) a combination of two or more of the above factors (Lees 1996a, Kletz 1986).

In the previous chapter we have reviewed some of the accidents which have occurred in different industries to illustrate the myriad ways in which human or equipment failure can cause an accident leading to loss of lives and property. A few more case studies are presented below.

At a refinery in France, a spillage occurred on 4 January 1996 when an operator was draining water from a 1200m pressurized propane sphere. The propane vapour spread over a radius of 150 m and was ignited by a car on the road. The pool of propane below the sphere engulfed the vessel in flames. The resultant boiling-liquid-expanding-vapour explosion (BLEVE) killed a fireman and 17 others. The conflagration took 48 hours to be controlled and caused extensive damage to the refinery.

On 22 June 1974, a 16-inch elbow of a pipe carrying potassium carbonate solution in a fertilizer plant at Tamil Nadu, India, ruptured suddenly, splashing the hot solution into the nearby control room. The toughened glass panes shattered; 8 people died in the control room instantly, one died in the hospital and another sustained grievous injuries.

On 8 March 1984, an explosion in a refinery at Kerala destroyed a fire tender along with the shed wherein it was housed, besides a chemical warehouse, cooling tower and other facilities. Later investigations revealed many shortcomings in the plant layout.

On 12 December 1987, a crude oil storage tank in a refinery at Maharashtra, India, started boiling over, spilling the contents on the dike around it. Emergency services and fire brigades, who were alerted, tried to evacuate the contents. After four hours of pumping out, the tank caught fire and exploded, spilling the contents. Eight hours of vigorous fire fighting had to be carried out before the fire could be controlled. There was extensive damage to the property. A larger dike and the provision of a separate dike for a large tank the one involved would have helped to prevent the spread of fire to other tanks.

On 5 November 1990, an explosion at the offside battery of

compressors at a gas-cracking plant in Maharashtra, India, killed 35 people, besides causing heavy damages to property and business interruption losses. Among the deficiencies in the layout, identified after the disaster, was the location of a contractor's shed dangerously close to the gas compressors. Less publicized, but perhaps of greater consequence, was the lack of facility to shut down the flow of hydrocarbon at the site itself. The plant personnel had to run to the control room faster than the vapour that followed them to close the feed valve.

An accident took place on 18 April 1989, in a 14-inch natural gas pipeline in a gas company in India. The pipeline was carrying compressed natural gas at a pressure of about 295–298 *psi* from the compressor station to various consumers. The spot of the accident was about 730 feet from the compressor station. A security personnel heard a loud sound at about 0950 hrs and saw a huge cloud of black smoke emanating from the ruptured pipeline which caught fire immediately. The flame rose as high as 150 feet during the initial stage.

The fire damaged buildings consisting of the general stores and the office of the materials department. Two employees died and six others received burn injuries. Investigations revealed that the portion of the pipeline which had blown off was extensively corroded as compared to other portions of the pipeline. The underground pipeline was close to the materials department where old lead cells were stored. The corrosion could have been due to the leakage of spent weak acid which seeped through the ground and corroded the buried pipeline.

The Siberian accident: Perhaps the most macabre accident, next only to the Bhopal gas tragedy in its severity, occurred on 3 June 1989, near Nizhnevartovsk in Western Siberia. Engineers stationed there noticed a sudden drop in pressure at the pumping end of an LPG pipeline. The pipeline was commissioned in 1985 to carry mixed LPG to feed the industrial city of Ufa. Instead of investigating the trouble, the engineers responded by increasing the pumping rate in order to maintain the required pressure in the pipeline. The actual leakage point was about 890 miles downstream between the towns of Asma and Ufa where the pipeline was installed about half a mile away to the side of the Trans-Siberian Railway. The smell of escaping gas was reported from the valley settlements in the area but no one did anything about it. The escaping, liquefied gas formed two large pockets in the low lying areas along the railway line. The gas cloud then drifted for a distance of 5 miles. Some hours later, after the main leakage had started, a train from Nizhnevartovsk destined for the Red Sea resort of Alder was approaching the leakage area when the driver noticed a fog in the area that had a strong smell. The driver of another train approaching from the opposite direction (Alder to Nizhnevartovsk) saw much the same as he approached the West bound train. Both trains were packed, with a total of 1168 people on board, and as they approached the area, the turbulence caused by them mixed up the LPG mist and vapour with the overlying air to form a flammable cloud. One or the other train ignited the cloud. Several explosions took place in quick succession followed by a ball of fire that was about 1 mile wide and which raced down the railroad tracks in both directions. Trees were flattened within a radius of two and a half miles of the epicentre of the explosions and windows were broken upto 8 miles away. The accident left 462 dead and 796 hospitalized with 70–80 per cent burn injuries.

PREVENTION AND CONTROL OF ACCIDENTS

Considering the serious and ever-increasing risk posed by the chemical process industries as elucidated above, it is necessary to conduct risk assessment and develop strategies to prevent accidents or, if preventive measures fail, to cushion their adverse impacts (Khan and Abbasi 1995 and 1996a).

As detailed in the previous chapter, such exercises in risk assessment would involve the following essential steps:

a) to identify vulnerable spots or 'high risk' points in an industry;
b) to simulate accidents and assess the damage they may cause;
c) to use the results of the previous step in identifying the priority areas where preventive measures need to be introduced;
d) to develop disaster management plans based on 'b' & 'c' above (Khan and Abbasi 1995, Greenberg and Crammer 1991).

It is evident that the step 'b' is very important in quantifying the risk posed. We have, therefore, developed a software package MAXCRED (Khan and Abbasi 1998). The package incorporates state-of-the-art mathematical models, including some developed or modified by the authors, for rapid quantitative and comprehensive risk assessment. A brief description of MAXCRED is presented here.

MAXCRED—THE PACKAGE

MAXCRED is a software package developed at Risk Assessment Division, Centre for Pollution Control and Energy Technology, Pondicherry, India. The package enables simulation of accidents and estimation of their damage potential. The software is written in C++ language and is compatible with DOS as well as Windows working environments. The software is operable on a personal computer and requires a minimum of 1 MB RAM and 1.5 MB ROM space.

The software has four main modules (options): scenario generation, consequence analysis, file and graphics. The scenario generation module enables the development of accident scenarios based on the properties of chemicals, operating conditions and likely ways of malfunctioning that would cause accidental releases. The consequence analysis module uses the previously developed accident scenarios and advanced models of thermodynamics, heat transfer and fluid dynamics to forecast the nature of the accidents, and their potentialities to cause damage. This module also enables the estimation of the damage radii and the probabilities of degrees of damage. The file option enables the user to handle input/output information and the graphics with ease. With the last-graphics option, visual scenarios of accidents are generated. All-in-all MAXCRED is a versatile tool for risk assessment and is envisaged to be self contained in the sense that it does not need other packages for data analysis or graphics support.

Here, MAXCRED has been used to visualize different accident scenarios in a typical chemical industry's storage unit. The industry is a part of a large industrial complex situated near Muzzafarnagar, Uttar Pradesh, India. The results have been represented in the terms of risk (damage probability) contour maps over the accident site to enable easy and swift assessment of maximum credibility scenario.

NATURE OF LIKELY ACCIDENTS

In general, an industry may have four types of accidents-namely (i) fire, (ii) explosion, (iii) toxic release and dispersion (Dow Chemical Company 1980), and (iv) a combination of (i), (ii) and (iii). These are briefly discussed below. The logical sequence and interdependency of such accidents that are likely to occur in an industry presented in figure 7.1.

Explosion

Explosions in the storage or process units can be categorized in three main groups, according to mode of occurrence and damage potential (Khan and Abbasi 1994, Kletz 1986, Mallikarjunan and Raghvan 1989, Pitersen 1990, Kayes 1985). These explosions are initiated either by the thermal stratification of the liquid and vapour or by such high explosion shock waves, which have sufficient strength to rupture the reaction or storage vessels, or conduits. An explosion may or may not be accompanied with fire; depending upon the type of explosion and the chemical involved in the explosion.

Boiling liquid expanding vapour cloud explosion (BLEVE)

BLEVE is a phenomenon which results from sudden release of gas or liquid stored at temperatures above their boiling points. At the vent or release point, a sudden decrease in pressure results in explosive vaporization of the stored material leading to a blast effect. The magnitude of BLEVE mainly depends on the material capacity and its rate of release.

Unconfined vapour cloud explosion (UVCE)

UVCE generally occurs when sufficient amount of flammable material (gas or liquid with a high vapour pressure) is released and mixes with air to form a flammable cloud such that the average concentration of the material in the cloud is higher than the lower limit of explosion (Lees 1996a, Pitersen 1990). The

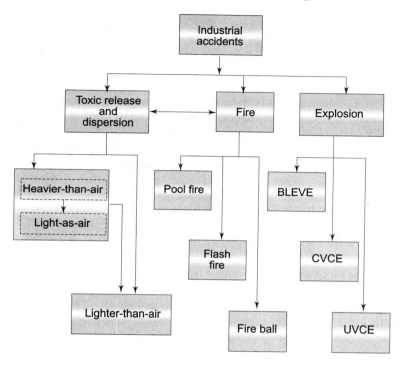

- Pool fire results from a slow yet continuous release of flammable liquid.

- Flash fire occurs due to the instantaneous release of material with a boiling point lower than atmospheric temperature.

- Fire ball is caused by a sudden combustion of vapour cloud of highly flammable chemicals.[17]

- Boiling liquid vapour cloud explosion takes place when there is sudden release of gas or liquefied gas stored at temperature above its boiling point; if the material is flammable there is a strong possibility of the formation of fire ball.[12,17]

- Confined vapour cloud explosion occurs in some sort of confinement; this is even more destructive than CVCE.

- Unconfined vapour cloud explosion is often more destructive than BLEVE.

Fig. 7.1 Tree diagram showing different accidental events and their inter-dependency

resulting explosion has a high potential of damage as it occurs in an open space covering large areas. The intensity of explosion mainly depends on the quantity of material released and the strength of the ignition source.

The explosive power of a UVCE can be expressed in terms of blast wave characteristics (overpressure, overpressure-impulse, reflected pressure, duration of shock wave etc). The peak overpressure is a very important parameter; its magnitude depends on the speed of flame propagation. Any obstruction in flame propagation enhances the blast effect.

Confined vapour cloud explosion (CVCE)

CVCE, as the name suggests, is a vapour explosion occurring in one or other type of confinement (Pitersen 1990, Kayes 1985). Explosions in vessels and pipes are examples of CVCE. Excessive generation of high pressure in the confinement leads to this type of explosion. It also has high potential for causing damage as it may generate fragments (missiles) propelled at high velocities which can cause more accidents. The energy delivered to the fragments by the blast wave causes the fragments to become air-borne and to act as missiles. The missiles are characterized by velocity, weight and penetration strength. However, the cumulative effect of CVCE depends upon the mass of material involved in the explosion and the explosion pressure.

Fire

Spillage of flammable material (liquid/gas) may lead to fire which could be triggered off by any of these ignition sources: an electric spark; a momentary flame due to welding operation; atmospheric friction or burning of a match stick. In case of highly flammable materials the fire may be started even by the mild friction caused by atmospheric disturbances (Khan and Abbasi 1986a, Kayes 1985, Contini et al. 1991). Generally the fire effects are limited to areas close to the source of fire (approx. 200m radius). However, industrial fires can have a greater pervasive effect. Industrial fires are mainly categorized into three groups, described below.

Pool fire: Continuous release of flammable liquid results in pool fire (Khan and Abbasi 1996a, Dow Chemical Company 1980,

Kayes 1985). The characteristics of such fires mainly depend on the duration of release, saturation pressure and flammable properties of materials.

Flash fire: Flash fire occurs mainly due to the instantaneous release of material having a boiling point lower than atmospheric temperature (Kayes 1985, Contini et al. 1991). It does not explode when the material release rate and flame speed are not high enough. However, it spreads quickly throughout the flammable zone of the vapour cloud.

Fire ball: An instantaneous ignition of flammable vapour cloud would lead to the formation of a fire ball (Khan and Abbasi 1996a, Greenberg and Crammer 1991, Kayes 1985, Contini et al. 1991). The radius of fire ball, its radiation heat intensity and temperature in the fire ball depend upon the dimension of the flammable cloud as well as the mass of the vapour released in the cloud. A very high temperature of 500 to 1500°K is developed in the confines of the fire ball. It is potentially the most disastrous of industrial fires that may be caused by highly flammable gases stored or processed under pressure.

The damage associated with such fires may be assessed on the basis of the dose of heat radiation received from them in a given time interval.

Toxic release and dispersion

Vapour clouds from industrial installations arise principally from the accidental release of gases, flashing liquefied gases or evaporation of spilled liquids. The toxic vapour (gas) cloud is likely to be dangerous even at much greater distances from the point of release than their flammable counterparts. This is mainly due to the ease of their dispersion and the high probability of coming in direct contact with living systems (Contini et al. 1991, Lees 1996b).

Release conditions: To estimate the characteristics of dispersion of gases due to an accidental release, the following accidental release conditions (with appropriate models) have been considered:

a) Gaseous release.

b) Liquid release (at atmospheric pressure). This condition can further be categorized as: (i) liquid with a boiling point above ambient temperature which is processed or stored at a temperature below its normal boiling point; (ii) liquid with a boiling point below ambient temperature which is processed or stored at low temperature and atmospheric pressure.

c) Two phase release (liquid under pressure). This condition can also be further categorized into two classes: i) liquid with normal boiling point above ambient temperature which is processed or stored under high pressure and temperature above its normal boiling point; ii) liquid with normal boiling point below ambient temperature which is processed or stored under high pressure and temperature above the normal boiling point.

Dispersion: Dispersion is primarily governed by two facts:

a) momentum of release

b) density of the gas relative to air.

As long as the momentum of the escaping gas is significant, the density factor does not become operative but as soon as the momentum dies down to a level where the ambient air movements could effect dispersion, the density factor takes over to influence the shape of the plume.

When the gas escapes at a high velocity as from a jet or a vent, the momentum effect is more prominent and lasts longer (due to higher velocity of release) than when the release velocity (venting velocity) is low.

According to Lees (1996a) release in the form of jets can be of four types: i) turbulent momentum jet in still air, ii) buoyant plume in still air; iii) plume dispersed by wind and iv) jet-turbulent plume dispersed by wind. The behaviour of such jets and vents is as relevant to the intended discharges as to accidental discharges. The behaviour of the dispersion of such jets depends on the relative importance of discharge momentum, buoyancy effects and of wind turbulence. We have used a turbulent jet model along with plume dispersion, as in modified plume path theory to estimate this mode of dispersion.

Once the gas loses its momentum, it is influenced by the density of the gas-air mixture relative to air. A difference in the molecular weight and/or in the temperature between the gas and the ambient air creates, in principle, such a density difference, but this density difference will affect the behaviour of the cloud only if the concentration of the gas is sufficiently high. A large proportion of the liquid droplets and a low air humidity favour the formation of a gas cloud heavier than air. A heavy cloud behaves differently from one of neutral density in several important aspects. It spreads not only downwind but also upwind, it is flatter in shape and the mechanisms of mixing with the air are different. Compared to dispersion of lighter-than-air or as-dense-as-air gases, heavy gas dispersion has been studied to a much lesser degree. Only a handful of models are available to handle heavy gas dispersion. We have adopted the Box model for heavy gas dispersion.

Dispersion of gases (gas-air mixture), whose density is equal to or less than that of air, under the influence of ambient air movements is characterized by neutral buoyancy dispersion. Even a heavy gas acquires a dispersion pattern akin to that of neutral buoyancy dispersion when density-driven turbulence becomes weak (in other words the density difference between an air-gas mixture and air becomes negligible) as more and more ambient air is entrained in the cloud causing atmospheric turbulence to dominate the dispersion process. We have used Gaussian plume and puff models for continuous and instantaneous release conditions respectively to predict the behaviour of neutral dispersion.

CASE STUDY

A risk assessment study has been carried out for a typical chemical industry situated in a congested industrial complex (figure 7.2) and engaged in manufacturing epichlorohydrin (EPI). EPI is a chemical which can cause fire as well as toxicity. Moreover, the manufacture of EPI involves the use of various hazardous chemicals such as propylene, chlorine and allylchloride at high temperatures and pressures. The industry and nearby areas are threatened by different types of hazards (explosion, fire, toxic gaseous release and corrosive liquid release).

Process summary

Purified propylene in the gaseous phase reacts with chlorine to give allylchloride. The crude allylchloride is purified by passing through a fractionating column. The purified chemical

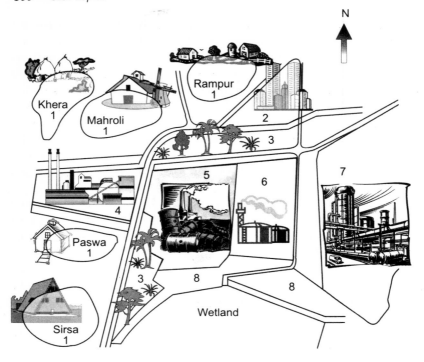

Key:

1. Populous villages

2. Residential buildings of Housing Board

3. Green belt

4. Fertilizer industry

5. Epichlorohydrin industry

6. Petrochemical industry

7. Petroleum refinery

8. Waste lands

Fig. 7.2 Layout of the study area showing location of industries and their surroundings

subsequently passes through the chlorohydration process at nearly ambient temperature to yield glycerol dichlorohydrin (GDH). GDH is further subjected to hydrolysis at 10°C using lime as a saponification agent. The crude EPI, thus produced by hydrolysis, passes through the fractionating column to yield pure EPI, which is then stored in tanks. The main units of an EPI plant are listed in table 7.1.

We have conducted a risk analysis of the complete plant. In order to optimize the time and effort, a hazard identification and ranking exercise was first carried out. The exercise is meant to identify the type of hazards present and a rough estimation of their hazard potential. The units are accordingly characterized and ranked in priority for a comprehensive study. Dow's Fire & Explosion Index has been used for this step and a summary of results thus obtained are presented in figure 7.3. It is clear from the figure that storage of propylene and chlorine are the most hazardous, followed by the storage of fuel oil and allylchloride. Other process units: chlorination, chlorohydration and quenching, are comparatively less hazardous. All these units

Table 7.1 The main units of an epichlorohydrin (EPI) plant

Units	Reference used in figure 7.2	Operation
Propylene purification	A	Fractionation
Propylene chlorination	B	Reaction
Allylchloride purification	C	Fractionation
Chlorine absorption in water	D	Absorption
Chlorohydrination	E	Reaction
Hydrolysis and EPI purification	F	Reaction & Distillation
Quenching	G	Heat transfer
Vent scrubbing	H	Separation
Propylene bullets	I	Storage
EPI tanks	J	Storage
Fuel oil tanks	K	Storage
Chlorinated organic tanks	L	Storage
Chlorine storage	M	Storage

have been further subjected to MAXCRED for hazard assessment and damage potential estimation.

For the sake of brevity, we give details of the study of only the most hazardous units in the plant (storage unit).

Table 7.2 Operating conditions of different storage units

Hazardous chemicals	No. of Tanks	Capacity of each tank (Ton/m³)	Operating pressure (atm)	Operating temperature (°C)
Propylene	2	60 T	15.20	40
Chlorine	2	100 m³	4.25	20
Allylchloride	2	50 m³	1.00	35
Epichlorohydrin	2	450 m³	2.25	40
Fuel oil	1	400 m³	1.00	40
Chlorohydrine	1	51 m³	1.22	40
Chlorinated organics	1	100 m³	1.00	40

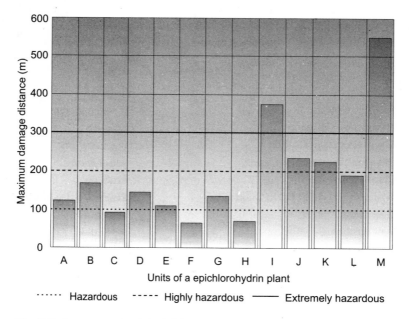

Fig. 7.3 Damage potential of different units in a epichlorohydrin plant (the details of A,B,C, ... etc. are provided in table 7.1)

The storage unit comprises the vessels storing various chemicals at high temperature and pressure (table 7.2). Each storage vessel (for each hazardous chemical) has been subjected to MAXCRED for hazard assessment and damage potential quantification. To felicitate understanding of the use of MAXCRED, the results of the study are presented in steps that follow the algorithm of MAXCRED.

Generation of accident scenarios

Based on the history of major accidents in process industries and the authors' experience, the following scenarios have been visualized for accidents in different storage units.

Scenario 1: An excessive pressure development in the storage vessel of propylene (under high pressure and temperature) leads to confined vapour cloud explosion (CVCE). The vapour cloud, generated by CVCE, on ignition turns into a fire ball and consequently damages the other storage vessels (chlorine, allylchloride, epichlorohydrin and fuel oil).

Scenario 2: A sudden release of pressurized chlorine triggers a boiling-liquid-expanding vapour explosion (BLEVE) and dispersion of toxic vapour cloud; in other words an explosive release followed by toxic dispersion.

Scenario 3: Unconfined vapour cloud explosion (UVCE) occurs in the epichlorohydrin unit accompanied by a pool fire. This will happen if a small leak in the storage vessel releases material at a moderate flow rate and forms a vapour cloud in the atmosphere which on ignition leads to UVCE. Due to the heat generated by UVCE or by an external source of ignition, the remaining material in the vessel or dike catches fire causing a pool fire. A UVCE followed by pool fire can damage neighbouring vessels, due to excess shock waves and heat load, and trigger secondary and higher order accidents.

Scenario 4: Instantaneous release is followed by pool fire in the allylchloride storage unit.

Scenario 5: There is a pool fire in the fuel-oil storage unit.

Scenarios 6 and 7: Toxic release of chlorine and chlorinated organic material is followed by evaporation and dispersion.

These scenarios have been processed for damage estimation through MAXCRED. A brief note on the damage-effect calculation models used for the detailed study is presented below.

Damage-effect calculations for the accident scenarios: The explosion, fires and toxic dispersions eventually cause damage in four ways. The potential of these effects can be expressed in terms of the probit function, which relates the percentage of people affected in a bounded region due to a particular accident event by a normal distribution function (Khan and Abbasi 1996a, Greenberg and Crammer 1991, Pitersen 1990, Contini et al. 1991, Lees 1996b).

1) *Heat radiation effect*

The probit function for 100 per cent lethality for heat radiation is given as:

$$Pr = -36.38 + 2.56 * \ln [t * q^{4/3}]$$

Where q is defined as thermal load (kw/m^2); t is time of exposure (s); and Pr is probit value.

2) *Toxic effect*

Lethality of a toxic load is expressed in terms of probit function as:

$$Pr = a + b \ln (C^n * t)$$

Where a, b and n are constants; C is concentration in ppm; and t is time of exposure (s). The values of the constants for different gases are available in literature.

3) *Pressure and shock wave effect*

The probit equation for likelihood of death due to shock wave (lung rupture) is given by

$$Pr = -77.1 + 6.91 * \ln P°$$

For injury, the equation is

$$Pr = -15.6 + 1.93 * \ln P°$$

Where P° is peak overpressure(N/m^2).

4) *Missile effect*

The probit function for fatality in human beings or damage to vessels is expressed as:

$$Pr = -17.50 + 5.30 * \ln S$$

Where, S is the kinetic energy of the missile (Joules).

HAZARD QUANTIFICATION

The results of the calculations for different accident scenarios are summarized below.

Table 7.3 presents the summary of calculation (output of MAXCRED) for scenario 7.1. The missiles generated by CVCE may hit nearby targets and can lead to secondary explosions or toxic releases. Ignition of the vapour cloud generated by CVCE may cause a fire ball and hence severe heat radiation effect. The shock wave generated due to CVCE can cause injury as well as second order accidents by seriously damaging other vessels. It has been estimated that shock waves with 50 per cent probability

Table 7.3 The output of MAXCRED for *scenario 1*

Parameters		Values
Distance from accident epicentre	(m) :	200
Explosion: CVCE		
Energy released during explosion	(kJ) :	1.733260e+8
Peak over-pressure	(kPa) :	5230.851562
Variation of over-pressure in air	(kPa/s) :	2339.458984
Shock velocity of air	(m/s) :	1978.708984
Duration of shock wave	(s) :	33.019657
Missile characteristics		
Initial velocity of fragment	(m/s) :	1098.601074
Kinetic energy of fragment	(kJ) :	7.860266e+05
Fragment velocity at study point	(m/s) :	984.562378
Peneration ability at study point		
Concrete structure	(m) :	0.975414
Brick structure	(m) :	0.992585
Steel structure	(m) :	0.094620
Fire: Fire ball		
Radius of the fire ball	(m) :	278.059906
Duration of the fire ball	(s) :	113.623726
Energy released by fire ball	(kJ) :	2.468228e+10
Radiation heat intensity	(kJ/m²) :	23570.539062

of causing injury would be observed over an area of 500 m radius. The heat radiation effect with 100 per cent probability of lethality would be observed over an area of 300 m radius and missile effects with 50 per cent chances of damage would be observed across 750 m radius.

The sudden release of chlorine and the consequent BLEVE as per scenario 2 would lead to severe shock waves, and toxic releases (table 7.4). In this scenario the damage potential of the shock waves is estimated to be comparatively lesser than that in scenario 1. A 100 per cent damage-causing shock wave would be operative over an area of 400 m radius. As chlorine is non-combustible, no heat radiation effects would be observed, but a build-up of lethal concentration would take place over an area of 1500 m radius.

Table 7.4 The output of MAXCRED for *scenario 2*

Parameters		Values
Distance from accident epicenter	(m)	: 200
Explosion: BLEVE		
Total energy released	(kJ)	: 5. 786538e + 05
Peak over-pressure	(kPa)	: 134. 8982
Variation of over-pressure in air	(kPa/s)	: 57. 98675
Shock velocity of air	(m/s)	: 343. 9556
Duration of shock wave	(s)	: 33. 48331
No missile effect		
Toxic release & dispersion		
Light gas dispersion characteristics		
Gaussian instantaneous model:		
Concentration at a distance of 200 m	(kg/m^3) :	1. 845254e − 04
Concentration at cloud axis	(kg/m^3) :	1. 345267e − 02
Value of source height	(m)	: 5. 000000
Puff characteristics:		
Puff concentration at centre of cloud	(kg/m^3) :	8. 169945e − 04
Concentration at cloud edges	(kg/m^3) :	8. 139317e − 04
Distance along downwind	(m)	: 200. 000000
Dosage at study point	(kg/m^3) :	0. 078989

An UVCE as per scenario 3 would give rise to heat radiation effects, shock waves and missile effects. In addition, there would be a secondary impact of burning of released material in the dike or vessel which would again lead to an additional heat load. In scenario 3, missile effect is not significant (table 7.5).

Table 7.5 The output of MAXCRED for *scenario 3*

Parameters			Values
Distance from accident epicentre	(m)	:	200
Explosion: UVCE			
Total energy released by explosion	(kJ)	:	6.628088e + 05
Peak over-pressure	(kPa)	:	115.6722
Variation of over-pressure in air	(kPa/s)	:	42.56866
Shock velocity of air	(m/s)	:	453.5449
Duration of shock wave	(s)	:	21.90854
Missile characteristics:			
Initial velocity of missile	(m/s)	:	876.561
Kinetic energy associated with missile	(kJ)	:	1988426.3750
Fragment velocity at study point	(m/s)	:	234.740
Penetration ability at study point :			
Concrete structure	(m)	:	0.2165
Brick structure	(m)	:	0.2456
Steel structure	(m)	:	0.0021
Fire: Pool fire:			
Instantaneous model			
Radius of the pool fire	(m)	:	5.000000
Burning area	(m^2)	:	78.53749
Burning rate	(kg/s)	:	38.42223
Heat intensity	(kJ/m^2)	:	1160.191
Toxic release & dispersion :			
Light gas dispersion characteristics			
Gaussian instantaneous model:			
Concentration at a distance of 200 m	(kg/m^3)	:	6.642933e – 07
Concentration at cloud axis	(kg/m^3)	:	5.335456e – 05

Contd.

Table 7.5 (Contd.)

Parameters	Values	
Puff characteristics:		
Puff concentration at centre of cloud	(kg/m^3) :	2.932708e–05
Concentration at cloud edges	(kg/m^3) :	2.972075e –05
Distance along downwind	(m) :	200.000000
Dosage at study point	(kg/m^3) :	0.0008844

However, shock waves of intensity high enough to damage all objects coming in their way would be persistent in an area of 300 m radius. The combined impact of heat radiation, UVCE, and pool fire would be lethal over an area of 600 m radius. Due to intense heat load, some of the chemical would evaporate and disperse, causing its build up to a lethal concentration over an area of 200 m radius.

The model results for scenario 4 are presented in table 7.6. Lethal heat load would be observed over the study area; the maximum damage distance would envelope an area of 400 m radius.

The output of MAXCRED for scenario 5 (fuel oil) is tabulated in table 7.7. Lethal heat load would be observed over an area of 350 m radius.

The results of scenarios 6 and 7 have been presented in table 7.8 and 7.9 respectively. In both cases the lethal concentration (100 per cent chances of fatality) is confined to an area of 200 m radius.

Table 7.6 The output of MAXCRED for *scenario 4*

Parameters	Values	
Distance from accident epicentre	(m) :	200
Fire: Pool fire		
Continuous model		
Burning area	(m^2) :	75.0000000
Burning rate	(kg/h) :	15042.8281
Heat intensity	(kJ/m^2) :	986.848602

Table 7.7 The output of MAXCRED for *scenario 5*

Parameters	Values		
Distance from accident epicentre	(m)	:	200
Fire: Pool fire			
Instantaneous model			
Radius of the pool fire	(m)	:	5. 000000
Burning area	(m^2)	:	78. 53749
Burning rate	(kg/s)	:	38. 42223
Heat intensity	(kJ/m^2)	:	1160. 191

Table 7.8 The output of MAXCRED for *scenario 6*

Parameters	Values		
Distance from accident epicentre	(m)	:	200
Fire: Pool fire			
Continuous model			
Burning area	(m^2)	:	75. 00000
Burning rate	(kg/h)	:	11842. 82
Heat intensity	(kJ/m^2)	:	336. 84860

Table 7.9 The output of MAXCRED for *scenario 7*

Parameters	Values		
Distance from accident epicentre	(m)	:	200
Fire: Pool fire			
Continuous model			
Burning area	(m^2)	:	75. 0000000
Burning rate	(kg/s)	:	13304. 3451
Heat intensity	(kJ/m^2)	:	456. 848602

Risk estimation

Using the results obtained by MAXCRED and the probability of occurrence of the accident scenarios, individual fatal risk factors have been estimated. The probability of occurrence of an individual event has been adopted from literature (Reliability Directorate 1992, European Community 1992, Khan and Abbasi 1996b). The risk factor is a direct representation of the threat (taking into consideration both damage potential and probability of occurrence) to an individual in an area.

In order to enable visualization of accident scenarios, their risk contours have been drawn over the study area (figures 7.4–7.8). It may be seen that the risk contours for scenarios 1–3 (figures 7.4–7.6) extend beyond the boundaries of the industry and envelope other industries and nearby populated areas. The risk contours for other scenarios are confined to the campus of the EPI plant but the risk contours for scenarios 4 and 5 extend upto other storage vessels in the factory (figures 7.7 and 7.8). This may cause secondary accidents, the impacts of which may go beyond the factory confines.

To summarize, scenario 2 represents the worst likely disaster within the realm of credibility. It has the largest area-of-lethal-impact (shock wave over an area of radius 500 m and lethal concentration across an area of radius 1500 m). Further, the most thickly populated areas (including residential areas of Mahroli and Rampur) lie within its range. If one considers the cumulative effects, scenario 1 would come out as the worst, as more intense impacts (in terms of heat radiation, shock waves and missiles) are observed per unit area in this scenario. Of the eight scenarios, scenario 1 is the most likely to cause cascading effects as missiles, shock waves and radiation effects would be generated simultaneously and other industries or units dealing with flammable and toxic materials are situated within the striking distance of the primary accident of scenario 1. Scenarios 3, 4 and 5 also have the potential to lead to secondary accidents as severe heat load generated in these would encompass other storage vessels. All-in-all, scenario 2 is the worst as far as primary effects are concerned whereas scenario 1 is the worst in terms of its potentiality of causing cascading (domino) effects.

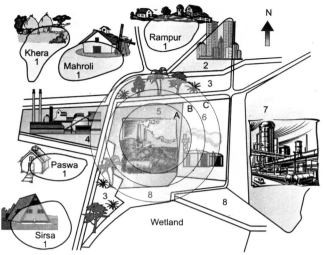

Fig.7.4 Risk contours indicating impact area for an accident in a propylene storage vessel (scenario 1), due to severe risk (A), high risk (B), and moderate risk (C). (Key: see Fig. 7.2)

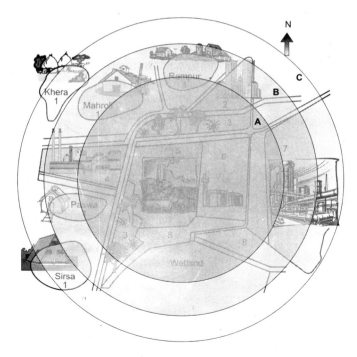

Fig.7.5 Risk contours indicating impact area for an accident in a chlorine storage vessel (scenario 2), due to severe risk (A), high risk (B), and moderate risk (C). (Key: see Fig. 7.2)

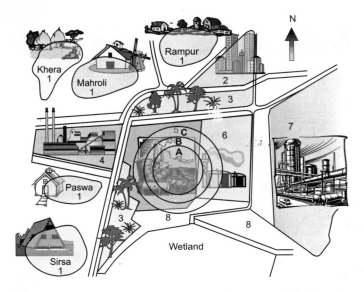

Fig. 7.6 Risk contours indicating impact area for an accident in an epichloro-hydrin storage vessel (scenario 3), due to severe risk (A), high risk (B), and moderate risk (C). (Key: see Fig. 7.2)

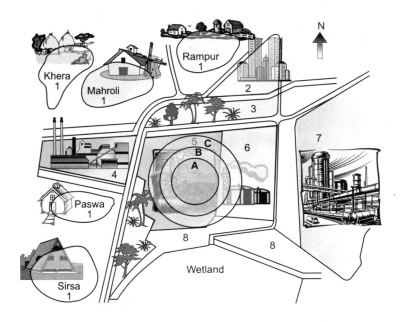

Fig. 7.7 Risk contours indicating impact area for an accident in an allychloride storage vessel (scenario 4), due to severe risk (A), high risk (B), and moderate risk (C). (Key: see Fig. 7.2)

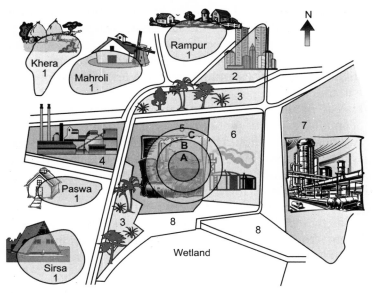

Fig. 7.8 Risk contours indicating impact area for an accident in a fuel oil storage vessel (scenario 5), due to severe risk (A), high risk (B), and moderate risk (C).(Key: see Fig. 7.2)

CONCLUSIONS

This chapter demonstrates the applicability of the software package MAXCRED for performing quantitative risk analysis. The package generates different credible accident scenarios, and quantifies the damage they can cause. This information can then be used in developing strategies for preventing accidents and to dampen their adverse impacts if the accidents do take place. The applicability has been illustrated with a case study of a typical chemical industry situated in Muzzafarnagar, Uttar Pradesh, India.

In the first step—hazard identification and ranking—storage and process units involving chlorination, chlorohydration and quenching were identified as the most vulnerable vis a vis propensity for causing accidents. A detailed study of the credible accidents in storage units and their impacts was then carried out with the help of MAXCRED. The studies reveal that the storage units of propylene, chlorine and epichlorohydrin are the most hazardous, and accidents in these units may cause severe damage to the factory and its surroundings.

To reduce the hazards associated with the storage of chlorine, propylene and epichlorohydrin, proper hazard minimization or mitigation measures should be taken. A few suggestions are made in this context:

   i) Instead of one or two large-capacity vessels several vessels of smaller capacity should be used for storage.

  ii) Adequate space should be kept between the storage vessels and buffers provided between them so that the adverse consequences of failure in one of them do not cause second or higher order accidents.

 iii) Sensitive gas detecting devices for flammable and toxic gases should be installed in the storage area and other units.

 iv) There should be regular and thorough inspections of the electronic control equipment followed by meticulous maintenance.

  v) Sufficient quantities of inert gases should be readily available to dilute the concentrations of toxic or flammable gases if they escape into the atmosphere, and to control fire.

 vi) A thorough emergency-preparedness strategy should always be kept in position, fortified by periodic drills or 'dry runs' so that damage is contained in the event of an accident.

REFERENCES

Abbasi, S.A. and V. Venilla (1994) 'Risk Assessment'. In *Encyclopedia of Environmental Pollution and Control*, edited by R.K. Trivedi. Karad: Enviromedia.

Contini, C.A., A. Amendola and I. Ziomas (1991) 'Benchmark Exercise on Major Hazard Analysis'. Amsterdam: Joint Research Centre–ISPRA.

Dow Chemical Company (1980) *Dow Fire and Explosion Index: Hazard Classification Guide*, 6th ed. Midland: Dow Chemical Company.

European Community (1992) 'Council Direction on the Major Accident Hazards of Certain Industrial Activities'. EEC Report No. 82/50/501. London: EEC.

Greenberg, H.R. and J.J. Crammer (1991) *Risk Assessment and Management for Chemical Process Industries*. New York: Van Nostrand Reinhold Publishers.

Kayes, P.J. (1985) *World Bank Manual of Industrial Hazard Assessment Techniques*. London: Technica Ltd.

Khan, F.I. and S.A. Abbasi (1994) 'Anatomy of Industrial Accidents'. In *Risk Assessment in Chemical Process Industries: Advanced Techniques*, edited by F.I. Khan and S.A. Abbasi. New Delhi: Discovery Publishing House.

—— (1995) 'Risk Analysis: An Optimum Scheme for Hazard Identification and Assessment'. Proceedings of National System Conference, Coimbatore, October 14–15.

—— (1996a) 'Major Accident Case Studies in Chemical Process Industries'. *Chemical Engineering World* (September), pp. 15–19.

—— (1996b) 'Accident Simulation in Chemical Process Industries Using Software MAXCRED'. *Indian Journal of Chemical Technology* 3: 339–44.

—— (1998) 'MAXCRED—A Software Package for Quantitative Risk Analysis'. *Environmental Software*, in press.

Kletz, T.A. (1986) *What Went Wrong: Case Histories of Process Plant Disasters*. London: Gulf Publication.

Lees, F.P. (1996a) *Loss Prevention in the Process Industries*. Vol. 1. London: Butterworths.

—— (1996b) *Loss Prevention in the Process Industries: Failure and Event Data*. Vol. 2. London: Butterworths.

Mallikarjunan, M.M. and K.V. Raghvan (1989) 'Risk Assessment Techniques and their Applications'. Proceedings of National Conference on Safety Systems, New Delhi, May 2–4.

Pitersen, C.M. (1990) 'Consequences of Accidental Release of Hazardous Material'. *Journal of Loss Prevention in Process Industries* 3: 136–41.

Reliability Directorate (1992) 'Failure Frequency of Hardware Components'. Report No. 87–2981R. 27/NVE. Washington D.C.: Reliability Directorate.

8

Water Treatment
Disinfection

The mercury poisoning that occurred at Minamata, Japan, during the 1950s killed hundreds of people and physically handicapped several thousand others. When methyl isocyanate gas leaked during the night of 3 December 1984, at Bhopal, it left tens of thousands dead or maimed. The toll made Bhopal gas tragedy the worst industrial disaster ever.

It has been reported that the waters of Jamuna which feed the water supply system of Delhi contain higher-than-permissible levels of toxic pesticides (*Down to Earth*, March 1997). This may well be true and may enhance the number of people who would eventually die of cancer or other equally painful diseases.

But all these figures on the extent of the impacts of chemical pollutants, pale into insignificance if we consider the number of human beings who either die, or suffer debility, due to *biological* pollutants present in water—protozoa, viruses, bacteria and certain species of microalgae. These pollutants, which cause diseases such as hepatitis, cholera, typhoid, gasteroenterities and dengue, kill or incapacitate millions of people across India every year. And such pollutants are present everywhere, be it a mega city, a small town or a mere hamlet.

The process in any water treatment scheme aimed at destroying the pathogenic organisms responsible for the above mentioned diseases is called *disinfection*. Considering the widespread presence and devastating impact of biological pollutants in water, disinfection is unarguably *the most important water treatment step* in any water supply scheme.

In any typical water treatment system disinfection follows screening, flocculation and filtration, its sole aim being destroying or inactivating pathogenic micro-organisms.[1,2] (These numbers refer to references listed at the end of the chapter.)

In small towns where elaborate water treatment facilities incorporating the steps mentioned above are not available, disinfection is the only treatment given to water prior to its supply.[2] In rural areas, disinfection is the only *possible* mode of treatment of well-water, that does not require any infrastructure and which can be done just by introducing a measured amount of disinfectant chemicals in the wells.

In this chapter we present an overview of disinfectants used for water treatment, their capabilities and weaknesses, and the R&D thrusts being made worldwide to develop less expensive, less harmful and more potent disinfectants.

CHLORINE AS DISINFECTANT

Chlorine is the most widely used disinfectant and accounts for over 90 per cent of all the chemical disinfectants used in water treatment.[2–4] It is used directly in the elemental form or as compounds, such as calcium or sodium hypochlorite, which release elemental chlorine on reaction with water. Chlorine kills bacteria by inhibiting the activity of their different critical enzymes. The efficiency in killing bacteria, ease of use and economy compared to other disinfectants are the reasons for the worldwide use of chlorine as a disinfectant.

In recent years, however, the use of chlorine has been questioned,[5–50] due to its failure to inactivate viruses and other disease-carrying micro-organisms.[5–13] The toxicity of chlorine as well as chlorinated compounds towards non-target organisms has also been reported.[14–50] Resistance of pathogenic organisms towards chlorine has also been observed.[7–9]

Odour and toxicity problems

Chlorine reacts with compounds; such as ammonia, amino acids, phenols etc, present in raw water, to produce chlorinated derivatives with a pungent smell.[5,6] Some of these compounds persist after chlorination and are toxic to non-target organisms.[14-50]

Ingols and Saffney[51] reported that chlorinated organic compounds accumulate in the fat tissues of fish, and are ingested in significant quantities by consumers of these fish. Similar reports have appeared on the toxic effects of chlorine and chlorinated organic compounds on a large variety of aquatic life forms.[14-35] The adverse effects of chlorine on the reproductive potential, development, growth and survival of various aquatic organisms, especially fishes have been discussed by various researchers.[14-35]

The mutagenic activity of chlorine and its derivatives, and genetic damage due to chlorine has also been reported.[36-45] Carlo and Mettlin[46] reported a significant positive correlation between pancreatic cancer of humans and trihalomethane levels in drinking water. These findings have been supported by Maugh[47], Robeck et al.[48] and Pike.[49] Pfaffenberger et al.[50] subjected serum cholesterol levels to statistical analysis against several factors, and found that the serum cholesterol levels related to levels of organically bound chlorine in drinking water.

Inadequate protection from viruses by chlorine

Early studies[51,52] involving purified virus preparations at controlled pH, contact time, temperature and virus titre indicated that 7–46 times free chlorine was required to inactivate the coxsackie virus as was required to destroy *Escherichia coli*. Reports[53-55] concerning coxsackie and polio viruses have shown that chlorine doses considered adequate for the destruction of indicator bacterial pathogens (0.2 ppm residual chlorine with a 10 minutes contact time at pH 7) do not completely inactivate enterovirus. Investigations on the viricidal effect of chlorination of effluent from a high-rate trickling filter plant seeded with known concentrations of echo 9 and polio 1 virus[56] have revealed that different strains of viruses show different rates of inactivation. Polio 1 virus is more resistant to inactivation than *E. coli* or echo 9 virus.

The frequency with which enteroviruses are being isolated from chlorinated plant effluents and reports of resistance of bacteria to chlorine inactivation are causing great concern.[7-13, 57-61] Chlorine apparently affects bacteria by inhibiting the activity of their critical enzymes. Enteroviruses carry no such enzymatic activity and different mechanisms of inactivation involving different considerations and different extents of impact may well be in force.

Bates et al.[8] succeeded in increasing the resistance of poliovirus Type I strain L.Sc to free chlorine through 10 successive sublethal exposures with residuals of 0.6 to 0.8 mg/L chlorine for 30 minutes. The increased resistance was found to be progressive over several cycles rather than a single-step change, which indicated an adaptive response by the viruses upon repeated exposure to sublethal chlorine doses. Stanley and Cannon[9] isolated cyanophages LPP-1 and LPP-2 from various wastewater treatment plant sites and showed that the freshly isolated viruses were more resistant to chlorination than stock laboratory strains. A 25 minute exposure of the viruses to a combined chlorine residual of 2.5 mg/L in secondary effluent did not achieve 99 per cent inactivation.

Malakhova[10] recovered enteroviruses from purified and chlorinated wastewaters and observed a mounting resistance by strain of *Staphylococcus aureus*, E. *coli*, *Pseudomonas aeruginosa*, *Aerobacter cloacae* and *Klebsiella pneumoniae* to repeated contact with successively higher doses of 0.005 and 0.001 per cent chloramine. The validity of traditional indicator groups for chlorinated primary effluents was questioned afresh by Berg et al.[11] who observed more than 5 logs inactivation of indigenous feacal or total coliforms or feacal streptococci simultaneously with the inactivation of less than 2 logs of the indigenous virus population. The absence of feacal coliforms in such effluents may not necessarily indicate the absence of potential hazard from viruses. Davis[12] observed that the coliform and feacal streptococcus groups, which were reduced 3 logs by the treatment of wastewater with 1 mg/L of total residual chlorine, were much more sensitive than other bacterial groups.

Perrine et al.,[13] found the cyst *Acanthamoeba polyphase* to be insensitive to concentrations normally employed in pool disinfection. Carberry and Stapleford[57] reported from a study of the water distribution system of Wilmington, Delavare, that maintenance of a chlorine residual was not sufficient to ensure bacteriological quality.

Peterson and Roberts[58] found that after treatment of an active Hepatitis A virus inoculum with 0.5, 1.0 and 1.5 mg/L of free residual chlorine, the inoculum induced hepatitis in 14, 8, and 10 per cent and seroconversion in 29, 33 and 10 per cent respectively, of test marmosets. Rosenzweig et al.,[59] demonstrated the resistance of fungal conidia to chlorination and showed that colonisation of a water distribution system by fungi could occur even in the presence of a chlorine residual of 0.4 to 0.5 mg/L. Several other reports have shown that chlorination has failed to inactivate viruses and bacteria in a number of cases.[60–65]

OTHER DISINFECTANTS

Several alternatives to chlorine have been tried. The disadvantage of the alternative disinfectants is their higher cost as compared to chlorine. Some disinfectants also suffer from difficulty in handling and application, impart of colour or odour to water, and are toxic to non-target organisms. The reported alternatives, their superiority over chlorine, if any, and their drawbacks are briefly reviewed.

Chlorine dioxide[66–73]

Chlorine dioxide is an unstable gas, and therefore has to be generated *in situ* by the reaction of chlorine with sodium chlorite.

Chlorine dioxide is a stronger oxidant than chlorine. Thus unlike chlorine, it does not combine with ammonia and most organic impurities before oxidizing them, which is advantageous in taste and odour control. Unlike chlorine, the bactericidal efficiency of chlorine dioxide is relatively unaffected at pH 6–10. Chlorine dioxide has the advantage of being a good sporicide but is uneconomical. It is also as toxic to non-target organisms as chlorine.

Fluorine[74]

Elemental fluorine is the strongest of all oxidizing agents. It is capable of oxidizing water, which renders its use for water disinfection unfeasible. Hydrofluoric acid and fluorides have been used as preservatives against moulds and for disinfection of certain viruses[27], but there are no reports with regard to the bactericidal effectiveness of these compounds in concentrations which are low enough to be safe to non-target organisms.

Bromine[75-78]

Weightwise, bromine is more expensive than chlorine. Its handling properties are generally comparable if not more favourable.

Bromine reacts with ammonia in water to form monobromamine and dibromamine, but does not form a stable tribromamine. Unlike monochloramine, monobromamine is a strong bactericide, almost as strong as free bromine. Because monobromamine is a strong bactericide, and tribromamine is not formed, there is no need to proceed to breakpoint bromination (providing bromine slightly in excess of the minimum amount required for disinfection). This fact makes bromine a competitor of chlorine for disinfection of swimming pools for which both elemental bromine and hypobromites have been tried. Bromine has a stronger tendency to form irreversible compounds and perhaps, substitution compounds of unknown physiological effects. This and the higher associated costs are the major factors deterring the use of bromine for purification of drinking water and wastewaters.

Iodine[70, 79-81]

The physiological response of man to the consumption of iodinated drinking water has been observed by Black et al.[79] No evidence of harmful effects was found a sizable population for several years.

As the tendency of iodine (I_2)to form objectionable organic substitution compounds is much less that that of chlorine, post-disinfection with iodine provides longer lasting protection against pathogens, and alleviates the problem of offensive tastes and odours often caused by chlorine.

The biocidal properties of iodine are significant.[70, 80-81] A dosage of 1.3 mg/L of I_2 kills *E. coli* within 30 seconds and represents the minimum concentration that devitalizes *Streptococcus fecalis* within 2 minutes.

Although potable water disinfection with iodine is more costly, it has a number of advantages over chlorination, particularly for post-disinfection purposes.

Iodine does not combine with ammonia to form iodamines, but rather oxidizes the ammonia. It also oxidizes phenols rather than combining with them. Thus, less iodine is usually required to obtain a free iodine residual. Another advantage is that I_2 and HIO (hypoiodous acid) are equally good disinfectants.

Iodine has been used successfully in swimming pools because its action is less dependent on pH, temperature, time of contact and nitrogenous impurities than chlorine.

The disadvantage of iodine is that it may impart a green to yellow-green colour to the water, depending on the concentrations of I_2 and $I^{(-1)}_3$ — both of which are coloured. Fortunately the concentration of I_2 in the normal pH range is low, as is the concentration of I_3, and HIO is a colourless compound.

Iodine tablets have been used by armies for emergency water disinfection. It has been found that they have the same disinfecting power as chlorine and also kill amoebic cysts, which chlorine does not. This along with the fact that iodine does not form iodoamines, makes it a good disinfectant on which more research efforts are required for exploring its use in potable water disinfection.

Ozone

Ozone(O_3) is a powerful oxidizing agent, from both thermodynamic and kinetic standpoints.[82] The immediate bactericidal properties of ozone are superior to those of chlorine and are to a great extent independent of pH[82,83]. It is used extensively in USA and Europe, for disinfection and concurrent removal of taste, odour and colour and metals like iron and manganese; but is hardly used in developing countries.[83]

The reactivity of ozone presents problems in the disinfection of water containing lots of organic matter or oxidizable inorganic impurities[82]. In these cases, though the overall effect of ozonation on the improvement of water quality may justify the use of ozone, an additional complication arises from the fact that the decomposition of ozone in water does not permit long-term protection against pathogenic regrowth, or recontamination, and makes post-chlorination of ozone-treated water mandatory.

Disinfection with O_3 has the advantage of being effective against some chlorine-resistant pathogens like cysts and virus forms.[84-92] The simultaneous removal of many other objectionable compounds present in the water-supplies enhances the usefulness of ozonation. Ozone, unlike chlorine, neither imparts offensive tastes and odours to water, nor does it usually produce toxic substances, such as chlorinated hydrocarbons, which can result from chlorination practices. Two major deterrents in the use of ozone for disinfection of public water-supplies have been the initial cost of ozonation equipment and the fact that it provides no residual protection against recontamination, and must therefore be followed by chlorination. Toxicity of ozone towards non-target organisms has also come to light.[93,94] Destruction of the lamellar epithelium of the gill and hormonal imbalance have been reported to occur in rainbow trout[93] and bluegill as a result of long-term ozone exposure.

Hydrogen peroxide

Although hydrogen peroxide is a strong oxidant like ozone and chlorine, it is a poor disinfectant.[95,96] Peroxide disinfection, at concentrations of 1.5–5 per cent takes a minimum of about 2–4 hours.

At these concentration levels, peroxide produces an evident taste that must be removed after the contact period. On the other hand, peroxide may be used to destroy the taste of chlorine. The inadequacy of hydrogen peroxide as a disinfectant is attributed to the ability of the enzyme catalase to combine with peroxide and decompose it to water and oxygen. Organisms capable of producing catalase are therefore, able to avoid oxidative destruction by hydrogen peroxide.

Permanganate[97-99]

Potassium permanganate is used in the treatment of potable waters for taste, odour control and for the removal of inorganic compounds, especially iron, manganese and hydrogen sulphide.[97-99] It is a powerful oxidizing agent and like ozone, reacts extensively with organic matter and certain inorganic species in water.

Permanganate exhibits noticeable disinfecting properties, although the death rates of E. *coli* are lower with permanganate than with ozone or chlorine in solutions of comparable strength.[97] Reduction in the number of viable micro-organisms in permanganate-treated water may also be effected by the subsequent removal of manganese dioxide by coagulation and sedimentation.

Disinfection incidental to the use of permanganate for general water purification lessens, and often eliminates the necessity for prechlorination or intermediate chlorination. Permanganate is more expensive than chlorine weightwise, but it shares with ozone the advantage over chlorine in not producing compounds with an offensive taste, odour or potential toxicity.

Heavy metals[78,100,101]

Silver, copper, mercury, cobalt and nickel exhibit bactericidal properties but silver is probably the only heavy metal with any reasonable degree of efficacy for this application.

Concentrations of silver ion as low as 15 mg/L, or 15 ppb, are sufficient to destroy most organisms if given sufficient time.[100] For most applications, higher concentrations and shorter times are employed. Silver ion has been found to be lethal to E. *coli* in concentrations from 0.005–0.5 ppm, with contact times from 2–24 hours, respectively.[74] Silver ion in the ppb range imparts no taste to water and is apparently nontoxic to humans at these low concentrations.[100] The relative safety of silver for its use as a disinfectant stems from the fact that most silver salts have very low solubility. Copper and mercury also exhibit significant oligodynamic properties. Copper ion, usually applied in the form of cupric sulphate, is frequently used as an algicide but is ineffective against spores. A further disadvantage of copper is that it tends to precipitate, and this decreases its effectiveness in slightly alkaline water. Mercury is highly toxic to human cells as well as micro-organisms, and cannot be applied to drinking water. One major drawback with silver, and other heavy metals, is their tendency to react with organic matter in water and wastes, which inhibits bactericidal action as a result of complex formation or precipitation of the metal. Besides, the treatment of water with silver is about 200 times costlier than that using chlorine.[78]

Acids and bases

Disinfection of water and wastes by deliberate acid or base addition is not a common practice, but is often incidental to other treatment processes, such as limesoda softening.[74,78] Furthermore, the effect of pH on the disinfecting efficiencies of certain chemical disinfectants can be pronounced,[74,102] and is an important variable in most water disinfection operations.

Thermal disinfection

Direct application of heat is one of the oldest and most reliable methods for water disinfection.[78] Nearly complete sterilization can usually be accomplished by boiling the water for a few minutes. But disinfecting large volumes of water by heating is not feasible for economic reasons, and is therefore not used for municipal water treatment.

Repeated rapid freezing and thawing usually results in bacterial death due to rapidly induced changes in osmotic pressure. This temperature-based technique of disinfection is also of little practical significance for water and wastewater treatment owing to high energy requirements.

Ultraviolet irradiation

Ultraviolet light kills a cell, retards its growth or changes its heredity by gene mutation.[103] The bactericidal effects of intense sunlight or artificial light are primarily due to UV or short wavelength irradiation. Low-pressure mercury vapour lamps which emit a narrow band of radiant energy at 2537 Angstroms have been used for disinfection in small-scale installations, but largely for reasons of economics, have not found significant application in municipal water or wastewater treatment.[78,104] The basic and applied aspects of UV-light disinfection of water and wastewater have been reviewed by Legan.[104] There is a spurt of studies on disinfection by UV-light due to its potential for causing quick and risk-free disinfection,[105–113] but unfavourable economics has limited the application of this technology on a large scale.

Gamma and X-irradiations

A radiation which is electromagnetic in nature with a very short wavelength, produces ionization reactions within cell molecules, leading to death of cells.[78,114] The interaction of gamma (γ) rays with water also produces unstable atoms, free radicals and other species which chemically react with organic molecules and cause secondary radiation effects in viable cells.

Radiation has a high degree of effectiveness against spores, viruses and other micro-organisms. It is uneconomical, and this coupled with the care required in applying this method for disinfection restricts its use to items of high value per unit weight, such as food and drugs.[78]

Ultrasonic frequency waves and electron beams

The bactericidal activity of ultrasonic frequency waves (UFW) was demonstrated by Horwood et al.[115] Subsequent studies have so far emphasized the impracticability of using UFW in its known mode of application in water disinfection owing to the high costs.[115–119]

The efficacy of high energy electron beam for sludge disinfection has also been explored.[120–122] In digested sludge, doses of 400 rads result in complete disinfection of viruses, with higher doses required for raw sludge.[121,122]

A fundamental breakthrough in the search for a viricidal physical property and the technique of its application is needed to make the disinfection by physical methods feasible in large scale treatment of water or wastewater.

Membrane techniques

Membrane-based reverse-osmosis techniques have recently emerged as methods of considerable potential in obtaining risk-free disinfection. For long, factors such as membrane fatigue and high energy requirements had prevented membrane-based methods from being used on a commercial scale for disinfection. These drawbacks have been overcome to the extent that membrane-based small-scale systems are able to compete with UV-based systems of similar scale (such as 'Aquaguard' units). Further breakthroughs in membrane technology may improve the economics of such systems and might make them viable on larger scales.

Miscellaneous disinfectants

Electrical fields ranging from 500-2500 V/m,[123] microwave radiation,[124] ultrasound,[125] anodic oxidation,[126] ferrates,[127,128] quarternary ammonium compounds,[129] iodine anion-exchangers[130] and lime[131] are amongst other disinfectants which have been tried without becoming economically feasible.

Comparative studies on disinfectants

The comparison of alternative disinfectants for both water and wastewater has lately received much attention.[132–139] For wastewater disinfection, pilot plant studies by Keswick et al.[132] indicate that bromine chloride (BrCl) is more effective than chlorine, and that the resulting residual from BrCl treatment is less stable. Chlorine dioxide was found to be a more effective disinfectant than chlorine against bacteria and viruses in wastewater.[133–135] Chlorine dioxide requirement of secondary effluent appeared to be more than the chlorine in the study done by Longley et al.[133] However, kinetics and the extent of demand were found to be same by Aieta et al.[135] Following both chlorine dioxide and chlorine treatment and subsequent dilution, total coliforms and feacal streptococci exhibited regrowth over several days.[134]

Kott et al.[136] reported *Salmonella typhimurium*, *S. fecelis* and feacal coliforms to be more sensitive to chlorine than to ozone when exposed in tap water spiked with ammonia or organic matter. The opposite behaviour was noted with Poliovirus Type I (Sabin) and coliphages. Economic comparisons between ozone and chlorination alternatives for drinking water treatment show ozone to be more expensive than chlorine, except when activated carbon treatment for dechlorination is needed.[137]

The performance of five wastewater treatment plants using chlorine, ozone or UV-light as disinfectants was reviewed and possible data correlations were discussed by Fluegge et al.[138] The only significant correlation appeared to be an inverse relationship between viral or coliform densities and ozone dosage at one plant. The disinfection of advanced wastewater treatment effluents by free or combined chlorine or by ozone was examined using samples spiked with Poliovirus Type I. Cost analysis suggested that combined residual chlorination was the most cost effective.[139] Several studies comparing the efficiency of chlorine to that of chlorine dioxide have been conducted.[140-143] Bench scale studies suggested that on a mass-dose basis, coliform inactivation by chlorine and chlorine dioxide are equivalent.[140] However, using native coliphage as an index, chlorine dioxide was found superior on the same basis.[141] Chlorine dioxide was more rapid than chlorine in coliform inactivation and disinfection was accomplished at lower final residual concentrations with the former agent.[140] Sorber and Longley[142] compared chlorine and chlorine dioxide as disinfectants in field studies, using a gravity flow contactor and observed that chlorine dioxide caused inactivation at lower final residual concentrations as compared to chlorine. Geisser et al.[143] studied the disinfection of combined sewer overflows using chlorine and chlorine dioxide, and developed regression equations incorporating RMS velocity gradient, total Kjeldahl nitrogen, 5 day BOD disinfectants dosage and contact time as independent variables to predict effluent quality. Using these models in an economic analysis it was concluded that chlorine, at short contact times and high dosages, was the preferred disinfectant.

In the disinfection of poultry packing plant effluents, it was found that a dosage of 10 mg/L chlorine and a 45 minute-contact time resulted in consistent negative analysis for *Salmonella*, while treatment with 30 mg/L ozone for 60 minutes resulted in 20 per cent of the samples examined being positive for *Salmonella*.[144] In potable water-treatment, chlorination in accordance with the Soviet standard resulted in water that was stable for at least 15 days, while treatment with various amounts of ozone resulted in an increase in coliform counts after 1–3 days.[145] The use of ozone and UV-light for the treatment of shellfish hatchery waste was investigated; both disinfectants achieved 'complete disinfection' of shellfish pathogens of the genera *Vibrio* and *Pseudomonas*[146].

The use of iodophors and chlorine in the inactivation of *Vibrio pavahaemolytious* was reported and it was found that prior growth conditions had little effect on the sensitivity of the organisms.[147] Gottardi[148,149] summarized equilibrium calculations on the nature of species in solution when iodine and iodo-compounds are used as disinfectants. It was noted that the most bactericidal species of iodine are present over a broader pH range than those of either chlorine or bromine.[148] Certain N-iodo compounds were proposed as investigative tools for the study of germicidal properties of hypoiodous acid in the absence of elemental iodine.[149] A combination of hydrogen peroxide (10 mg/L), ascorbic acid (10 mg/L) and Cu^{2+} ions (0.5 mg/L) was found to cause 99 per cent reduction in the initial plate count and 99–99.9 per cent reduction in enterobacteriaceae after a 60 minute contact time.[150] The observed effects were attributed to the toxicity of Cu^{2+} – ascorbic acid complex whose penetration into the cell was promoted by hydrogen peroxide.

Amongst the various possible alternatives to chlorine as disinfectants, chlorine dioxide[151–154] and ozone[154–164] have received more attention.

CONCLUDING REMARKS

Although chlorine is widely used in the disinfection of water and wastewater throughout the world, problems arising from its use are sharply coming into focus. Chlorine residuals have been found to be toxic to non-target organisms and a majority of them are suspected to be carcinogenic for humans. Removal of these residuals from potable water prior to supply (dechlorination) renders the water prone to recontamination and also increases the cost of chlorine disinfection. The capability of chlorine to inactivate viruses has become questionable and instances are being reported throughout the world of the developing resistance towards chlorine of the enteric bacteria which have so far been believed to be completely inactivated by chlorine.

Some of the alternative disinfectants reported so far are capable of inactivating pathogenic organisms over the expected variation in water and wastewater temperature, quantities and composition, but at a prohibitively high cost. Ozone, which is being used on a large scale in developed countries, is not only more costly than chlorine but provides no residual disinfection, leaving the ozone-treated water prone to recontamination.

In view of the fundamental relation between disinfection and human health, and of the global nature of the problem, there is a need to search for disinfectants and develop disinfection techniques which would ideally have all advantages of chlorine disinfection and the additional advantages of: i) greater range and strength of disinfecting ability; particularly the effectiveness against the enteric viruses which survive chlorine treatment; ii) less adverse influence than chlorine on the taste and odour of water; iii) and non-toxicity towards non-target organisms.

REFERENCES

1. Holum, J.R. (1977) *Topics and Terms in Environmental Problems.* New York: John Wiley and Sons.

2. Hussain, S.K. (1976) *Text Book of Water Supply and Sanitary Engineering.* New Delhi: Oxford & IBH.

3. Morris, J.C. (1971) *Journal of the American Water Works Association* 63.

4. Haas, C.N. and J.P. Gould (1979) *Journal of Water Pollution Control Federation* 51.

5. Jolley, R.L., P.K. Roy and L. Thomas (1976) In *Proceedings of the Conference on the Environmental Impact of Water Chlorination,* edited by R.L. Jolly. London: Springfield.

6. Glaze, W.H., W.W. Robbert and P.M. Sillen (1976) Cited in *Proceedings of the Conference on the Environmental Impact of Water Chlorination,* edited by R.L. Jolly. London: Springfield.

7. Erickson, J.J. and H.R. Foulk (1980) *Journal of Water Pollution Control Federation* 52.

8. Bates, R.C., U.L. John and P. Sheffield (1977) *Applied Environmental Microbiology* 34.

9. Stanley, J.L. and R.E. Cannon (1977) *Journal of Water Pollution Control Federation* 49.

10. Malakhova, T.S. (1977) *Microbiology Abstracts* 12.

11. Berg, G., V.K. Talwin and M. Feig (1978) *Applied Environmental Microbiology* 36.

12. Davis, E.M. (1977) *Bulletin of Sanitation,* Panama, 82.

13. Perrine, D., K. Vukink and P. Sanie (1980) *Chemical Abstracts* 93.

14. Yosha, S.F. and G.M. Choen (1979) *Bulletin of Environmental Contamination and Toxicology* 21.

15. Thatcher, T.O. (1979) *Bulletin of Environmental Contamination and Toxicology* 21.

16. Dinnel, P.A., N. Fenix and P.K. John (1979) *Bulletin of Environmental Contamination and Toxicology* 22.

17. Seegert, G.L., Z. Dora and Q.R. Philip (1979) *Transactions of the American Fisheries Society* 108.

18. Stewart, M.E., R.T. Bodkins and A.M. Roy (1979) *Marine Pollution Bulletin* 10(6).

19. Liden, L.H., R.L. Roberts and V.V. Sanks (1980) *Journal of Water Pollution Control Federation* 53.

20. Tsai, G.F. and J.A. Mckee (1980) *Transactions of the American Fisheries Society* 109.

21. Rosehoom, D.P. and D.L. Richey (1980) 'Acute Toxicity of Residual Chlorine and Ammonia to Some Native Illinois Fishes'. National Technical Information Service No PB 80, 510771, USA.

22. Hammond, B. and J. Bishop (1980) 'Maximum Utilization of Water Resources in a Planned Community: Chlorine and Ozone Toxicity Evaluation'. National Technical Information Service, No PB 80, 131121, USA.

23. Ward, R.W., P.M. Ripple and K.K. Mathews (1980) *Water Resources Bulletin* 16.

24. Elmore, M.B., A.N. Joy and P.S. Marshall (1980) *Bulletin of Environmental Contamination and Toxicology* 24.

25. Scott, G.I., V.V. Jacobs and R.S. Davidson (1980) *Water Chlorination: Environmental Impacts and Health Effects* 3.

26. Hillman, R.E., N.S. Allana and P.T. Simon (1980) *Water Chlorination: Environmental Impacts and Health Effects* 3.

27. Burton, D.T., S.H. Rustom and N.A. Wytin (1980) *Water Chlorination: Environmental Impacts and Health Effects* 3.

28. Khalanski, M. and F. Border (1980) *Water Chlorination : Environmental Impacts and Health Effects* 3.

29. Laird, C.E. and M.H. Roberts (1980) *Water Chlorination: Environmental Impacts and Health Effects* 3.

30. Miller, W.R., S.T. Weinex and P.A. Naiz (1980) *Water Chlorination: Environmental Impacts and Health Effects* 3.

31. Trabalka, J.R., S.N. Swieta and Z.H. Yuneit (1980) *Water Chlorination: Environmental Impacts and Health Effects* 3.

32. Turner, A., and T.A. Thayer (1980) *Water Chlorination: Environmental Impacts and Health Effects* 3.

33. Sanders, J.G., and J.H. Ryther (1980) *Water Chlorination: Environmental Impacts and Health Effects* 3.

34. Murrary, S.A. (1980) *Water Chlorination: Environmental Impacts and Health Effects* 3.

35. Moore, G.S., R.P. Thayer and R.G. Krutchkoff (1980) *Journal of Environmental Science & Health (A)* 15.

36. Rapson, W.H., V.K. Heynes and T.T. Soong (1980) *Bulletin of Environmental Contamination and Toxicology* 24.

37. Moore, R.L., W.R. Lynn and D.P. Loucks (1980) *Water Research* 14.

38. Fallon, R.D. and C.B. Fliermans (1980) *Chemosphere* 9.

39. Cheh, A.M., S.T. Yeik and N.A. Fein (1980) *Water Chlorination: Environmental Impacts and Health Effects* 3.

40. Bempong, M.A. and F.E. Scully (1980) *Water Chlorination: Environmental Impacts and Health Effects* 3.

41. Maddock, M.B. and J.J. Kelly (1980) *Water Chlorination: Environmental Impacts Health Effects* 3.

42. Payne, J.E., N.A. Keyne and J.N. Jein (1980) *Water Chlorination: Environmental Impacts and Health Effects* 3.

43. Nazar, M. (1982) *Environmental Mutagenesis* 4.

44. Suessmuth, R. (1982) *Mutation Research* 105.

45. Clark, R.R. and J.B. Johnston (1982) *Influence of Chlorination and the Distribution System on Mutagens in a Potable Water Supply.* Report UTLU–WRC–82–0168, Government Report Announcements, USA.

46. Carlo, G.L. and L.J. Mettlin (1980) *American Journal of Public Health* 70.

47. Maugh, T.H. (1981) *Science Wash* 211.

48. Robeck, C.G., J.P. Martin, P.N. Saunders and R.L. Rutgers (1981) *Science of the Total Environment* 18.

49. Pike, M.C. (1980) *Water Chlorination: Environmental Impacts and Health Effects* 3.

50. Pfaffenberger, C.D., P.R. Jaffene, N.N. Loyla and S.A. Smiths (1980) *Water Chlorination: Environmental Impacts and Health Effects* 3.

51. Ingols, R.S. and P.E. Saffney (1964) 'Biological Studies of Halophenols'. Oklahoma Industrial Waste Conference, Oklahoma State University, 17 November.

52. Clarke, N.A. and P.W. Kabler (1954) *American Journal of Hygiene* 59.

53. Kelly, S. and W.W. Sanderson (1957) *Science* 126.

54. Kelly S. and W.W. Sanderson (1958) *American Journal of Public Health* 48.

55. Kelly S. and W.W. Sanderson (1960) *American Journal of Public Health* 50.

56. Shuval, H.I., S. Cymbalista, A. Wachs, Y. Zohar and N. Golbum (1966) *Advances in Water Pollution Research* 2.

57. Carberry, J.B. and L.R. Stapleford (1979) *Journal of the American Water Works Association* 71.

58. Paterson, D.A. and B.P. Roberts (1983) *Applied Environmental Microbiology* 45.

59. Rosenzweig, W.D., J.S. Simen and V.M. Meekins (1983) *Applied Environmental Microbiology* 45.

60. Hejkal, T.W., V. Tietjen and P.S. Rothmans (1979) *Applied Environmental Microbiology* 38.

61. Gerba, C.P. and C.H. Staff (1979) *Journal of Water Pollution Control Federation* 51.

62. Haman, S. and D.J. Milkowska (1980) *Roczeniki Panstwcz Zaklitsche Higene* 31(1).

63. Seyfried, P.L. and D.J. Fraser (1980) *Canadian Journal of Microbiology* 26.

64. Alverez, M.E. and R.T. O'Brien (1982) *Applied Environmental Microbiology* 43.

65. Nagy, L.A. and B.H. Olson (1982) *Canadian Journal of Microbiology* 28.

66. Benarde, M.A., G.G. Logie and G. Wood (1965) *Applied Microbiology* 13.

67. Benarde, M.A., G.G. Logie and G. Wood (1967) *Applied Microbiology* 15.

68. Walters, G.E. (1976) 'Chlorine Dioxide and Chlorine: Comparative Disinfection', M.S. thesis, John Hopkins University.

69. Rauh, J.S. (1979) *Journal of Environmental Science* 22(2).

70. Alverez, M.E. and R.T. O'Brien (1982) *Applied Environmental Microbiology* 44.

71. Aieta, E.M., J.D. Berg, P.V. Roberts and R.C. Copper (1980) *Journal of Water Pollution Control Federation* 52.

72. Roller, S.D., D.S. Coaster and Y.B. Sharman (1980) *Water Research* 14.

73. Longly, K.E. (1982) *Journal of Water Pollution Control Federation* 54.

74. Sykes, G. (1958) *Disinfection and Sterilization.* New York: Van Nostrand Reinhold Publishers.

75. Keswick, B.H., H.H. Chadwick and L.M. Peters(1978) *Government Report on Announcement of Index (US)*, 78.

76. Floyd, R., A.S. Patterson and C.C. Clay (1978) *Environmental Science and Technology* 12.

77. Keswick, B.H., V.V. Richards and R.M. Binny (1978) *Journal of the American Water Works Association* 70.

78. Crawford, H.B. and D.N. Fishcheh (eds) (1971) *Water Quality and Treatment*, American Water Works Association. New York: McGraw Hill, pp. 160–226.

79. Black, A.P., R.C. Kinman, W.C. Thomas, G. Greund and E.D. Bird (1965) *Journal of the American Water Works Association* 57.

80. Black, A.P., R.C. Kinman, W.P. Bonner, M.A. Kerin and A.A. Jabero (1968) *Journal of the American Water Works Association* 60.

81. Cramer, W.N., S. Connery, R.G. Moore and I.F. Fleming (1976) *Journal of Water Pollution Control Federation* 48.

82. Evans, F.L. (1975) *Ozone in Water and Wastewater Treatment*. Michigan: Ann Arbor Science.

83. Katzenelson, E., B. Kletter, H. Schechter and H.I. Shuval (1975) In *Chemistry of Water Supply, Treatment and Distribution*, edited by A.J. Rubin. Michigan: Ann Arbor Science.

84. Katzenelson, E. and N. Biedermann (1976) *Water Research* 10.

85. Perrich, J., L.F. Saunders and K.G. Barrington (1978) *Chemical Abstracts* 88.

86. Evison, L.M. (1978) *Progress in Water Technology* 10.

87. Lacker, D.S. and T. Lockowitz (1978) *Chemical Abstracts* 88.

88. Sproul, O.J., R.P. Thayer and W.J. Padgett (1979) 'Effect of Particulates on Ozone Disinfection of Bacteria and Viruses in Water'. USA EPA 600/2-79-089, USA.

89. Foster, D.M., M.D. Flood and K.K. Kuhn (1980) *Journal of Water Pollution Control Federation* 52.

90. Walsh, D.S., T.S.Goodenough, P.R. Roy (1980) *Journal of Environmental Science* 196.

91. Roy, D. (1982) *Journal of the American Water Works Association* 74.

92. Sproul, O.J. (1982) *Journal of Water Pollution Control* 14.

93. Wedemeyer, J., M. Rogers and K.C. McDonald (1979) *Journal of Fisheries Research Board*, Canada, 36.

94. Paller, M.H. and R.C. Heidinger (1980) *Environmental Pollution (Ser A)*, p. 229.

95. Schumn, W.C., C.N. Statterfield and R.Z. Wentoworth (1955) *Hydrogen Peroxide*. New York: Van Nostrand Reinhold Publishing.

96. Poffee, R., M.S. Searjent and B.J. Moffit (1978) *Zentralbl Bakteriol Paranitenpd Infekt Ionster Hyg I Abt* 166.

97. Cleasby, J.L., E.R. Baumann and C.M. Black (1964) *Journal of the American Water Works Association* 56.

98. Reidies, A.H. (1977) *Water Works Engineering* 116.

99. Peskator, H. (1979) *Water Sewage Works* 126.

100. Clark, J.W. and W. Vigesman (Jr) (1965) *Water Supply and Pollution Control*. Scranton, Pa: International Textbook Company.

101. Fair, C.M., J.C. Geyer and D.A. Okun (1968) *Water and Waste Engineering*. New York: John Wiley and Sons.

102. Engelbrecht, R.S., J.J. Winefordner and G.D. Schaumberg (1980) *Applied Environmental Microbiology* 40.

103. Oliver, B.G. and J.H. Carey (1976) *Journal of Water Pollution Federation* 48.

104. Legan, R.W. (1980) *Water Sewage Works* 127.

105. Antopol, S.C. and P.D. Ellner (1979) *Applied Environmental Microbiology* 38.

106. Hass, C.N. and G.P. Sakellaropoulos (1979) 'Proceedings of the ASCE Environmental Engineering Division Speciality Conference', p. 540.

107. Wolf, H.W., R.M. Petheric and K.D. Lilly Worth (1979) *Chemical Abstracts* 91.

108. Wolf, H.W., R.M. Petherick and K.D. Lilly Worth (1979) 'Utility of UV Disinfection of Secondary Effluent', In *Progress in Waste Water Disinfection Technology*, edited by A.D. Venosa. US EPA 600/9–79–018, p. 502.

109. Johnson, J.D., N. Levinson and F. Kozin (1979) 'Ultraviolet Disinfection of Secondary Effluent'. In *Progress in Wastewater Disinfection Technology*, edited by A.D. Venosa. US EPA 600/9–79–018, p. 509.

110. Scheible, O.K., N.A. Allana and Y.C. Meikle (1979) 'Full Scale Evaluation of UV Disinfection of a Secondary Effluent'. In *Progress in Wastewater Disinfection Technology*, edited by A.D. Venosa. US EPA, 600/9–79–D18, p. 517.

111. Bayliss, C.E. and W.M. Waites (1980) *Journal of Applied Bacteriology* 48.

112. Severin, B.F. (1980) *Journal of Water Pollution Control Federation* 52.

113. Peterasek, A.C., A.N. Andrek and P.K. Sshalz (1980) *Ultraviolet Disinfection of Municipal Wastewater Effluents.* US EPA 600/2–80–102.

114. Thomas, F.C. (1982) *Applied Environmental Microbiology* 43.

115. Horwood, M.P., J.P. Horton and V.A. Michm (1961) *Journal of the American Water Works Association* 43.

116. Horton, J.P., P. Murrey, J. L. Horwood and D.E. Phinney (1952) *Sewage and Industrial Wastes* 24.

117. Falkovskaje, L.J. (1956) *Hygiene and Sanitation* 21.

118. Faber, H.A. (1961) 'Proceedings of International Water Supply Congress and Exhibition', Berlin, 30 May–2 June.

119. *Water Science and Technology.* Special Issue, edited by D. Jenkins (1982). Oxford: Pergamon Press.

120. Levaillant, C. and C.L. Gallien (1979) *Radiation Physics and Chemistry* 14.

121. Metcalf, T.G. (1979) *Disinfection of Enteric Viruses in Sludge by Energized Electrons.* Report NTIS PB 80–104086.

122. Trump, J.G., W.J. Padgett and G. Schultz (1979) *Proceedings of the National Conference on Municipal Sludge Management* 8.

123. Smirnova, L.F. (1979) *Zhur Praktikle Khimistry.* Leningrad 13.

124. Iskandar, I.K., P. Tsokos and N. Levinson (1980) 'Government Report on Announcement of Index (US)'. Report SR–80–1, 80: 2633

125. Boucher, R.M.G. (1979) *Canadian Journal of Pharmaceutical Sciences* 14.

126. Krueger, D. (1980) *Pharmaceutical Industry* 42.

127. Waite, T.D. (1979) *Proceedings of the American Society of Civil Engineers Environmental Engineering Division* 105.

128. Schnink, T. and T.D. Waite (1980) *Water Research* 14.

129. Walfish, I.H. and G.E. Janauer (1979) *Water Air Soil Pollution* 12.

130. Suzuki, T. and K.T. Fan (1979) *Journal of Fermentation Technology,* Japan, 57.

131. Polprasert, C. (1981) *Water Research* 15.

132. Keswick, B.H., D.P. Loucks and W.R. Lynn (1980) *Journal of Water Pollution Control Federation* 52.

133. Longley, K.E., N.A. Thayer and E.B. Phelps (1980) *Journal of Water Pollution Control Federation* 52.

134. Berg, J.D., T.N. Sozoka and H.T. Vivek (1980) *Water Chlorination: Environmental Impacts and Health Effects* 3.

135. Aieta, E.M. P.Y. Heign, and S.N. Guinea (1980) *Water Chlorination: Environmental Impacts and Health Effects* 3.

136. Kott, Y., H.N. Tsoko and T.K. Peter (1980) *Water Chlorination Environmental Impacts and Health Effects* 3.

137. Dahi, E. (1979) *Water Chlorination: Environmental Impacts and Health Effects* 2.

138. Fluegge, R.A., P.N. Saiga and S.T. Huige (1979) In *Progress in Wastewater Disinfection Technology*, edited by A.D. Venosa. US EPA, EPA 600/9 79–018.

139. Dryden, F.D., J.S. Paul and U.P. Wydenx (1979) *Journal of Water Pollution Control Federation* 51.

140. Aieta, E.M., Y.Z. Peter and N.P. Reneix (1979) In *Progress in Wastewater Disinfection Technology*, edited by A.D. Venosa. US EPA, EPA 600/9 79–018.

141. Berg, J.D., H.N. Feig, and T.P. Geisser (1979) In *Progress in Wastewater Disinfection Technology*, edited by A.D. Venosa. US EPA, EPA 600/9 79–018.

142. Sorber, C.A. and K.E. Longley (1979) 'Proceedings of the ASCE Environmental Division Speciality Conference', p. 15.

143. Geisser, D.F., E.A. Coddington and K.N. Schulz (1979) *Journal of Water Pollution Control Federation* 51.

144. John, P.K. (1978) *Disinfection of Poultry Packing Plant Effluent Containing Salmonella*. Dearborn Chemical Co. Ltd., Environmental Protection Service, Canada. Econ Tech Rev Rep EPA 3–WP–78–9.

145. Rusanova, N.A, P.T. Casona and Y.T. Pzuoza (1979) *Georgian Sanitation* 4.

146. Blogoslawski, W.J., S.H. Neilsozki and P.N. Feig (1978) 'Proceedings of the Annual Meet, World Mariculture Society', 9.

147. Gray, R.J.H. and D.H. Hsu (1979) *Journal of Food Science* 44.

148. Gottardi, W. (1978) *Zentrobl Bakteriology Abstracts I* 167.

149. Gottardi, W. (1978) *Zentrobl Bakteriology Abstracts I* 167.

150. Ragab-Depre, N.J. (1982) *Applied Environmental Microbiology* 44.

151. Hubbs, S.A., P.T. Geig and J. Sanga (1981) *Journal of the American Water Works Association* 73.

152. Monscvitz, J.T. and D.J. Rexing (1981) *Journal of the American Water Works Association* 73.

153. Brett, R.W. and J.W. Ridgeway (1981) *Journal of Institution on Water Engineering Science* 35.

154. Fiessinger, F. (1981) *Science of the Total Environment* 18.

155. Chaspal, P. (1981) *Aqua* 4.

156. Nebel, C. (1981) *Public Works* 112.

157. Stoebner, R.A. and D.A. Rollag (1981) *Aqua* 4.

158. Roy, D., D.L. Roy and N.A. Roshan (1981) *Water Science Technology* 15.

159. Sibnoy, J. (1981) *Aqua* 5.

160. Clark, R.M. (1981) *Journal of the American Water Works Association* 73.

161. Roy, D. and T.G. Bergman (1982) *Water Research* 16.

162. Emerson, M.A. and F. Stolle (1982) *Applied Environmental Microbiology* 43.

163. Van Hoof, F. (1982) *Aquatic Science Technical Reviews*. London, 5.

164. Burleson, G.R. and T.M. Chambers (1982) *Environmental Mutagenesis* 4.

Index